PSAM 12

Probabilistic Safety Assessment and Management
22–27 June 2014 • Sheraton Waikiki, Honolulu, Hawaii, USA

O'ahu

CONFERENCE PROCEEDINGS

Volume 13 - Friday AM

PSAM 12

Probabilistic Safety Assessment and Management

22 - 27 June, 2014

Sheraton Waikiki, Honolulu, Hawaii USA

CONFERENCE PROCEEDINGS

Volume 13

Friday AM

Foreword

It is was our honor to welcome you to Honolulu, Hawaii, for the twelfth rendition of the Probabilistic Safety Assessment and Management (PSAM) Conference. The planning for PSAM Honolulu began back in 2007 (before PSAM 9 in Hong Kong), when we looked at several locations around the United States, included Arizona, California, Boston, and even considered locations in Oceania. Based upon the feedback both during and after the conference, PSAM 12 proved to be a great success.

We would like to thank all of the volunteers, those that served before, during, and after the Conference. Members of the Technical Program Committee, the Organizing Committee, the session chairs, and the presenters have our gratitude for making PSAM 12 the most memorable PSAM yet.

This publication represents the technical proceedings for the Conference. Due to the large number of published papers (a total of 391), we have subdivided the technical content (papers) into five volumes, one for each day of the conference.

On behalf of the International Association for Probabilistic Safety Assessment and Management Board of Directors, we hope that this publication will provide a valuable technical resource in addition to a reminder of the memorable stay in the Hawaiian Islands.

Dr. Curtis Smith
Technical Program Chairs

Dr. Todd Paulos
General Chair

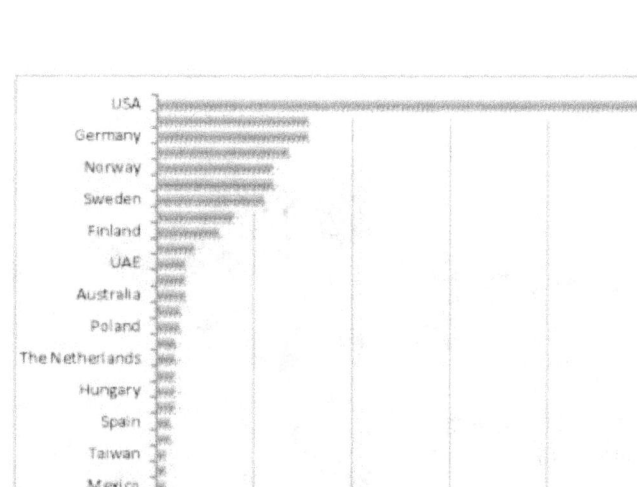

Number of Papers Presented at PSAM 12 (by country)

Sponsors

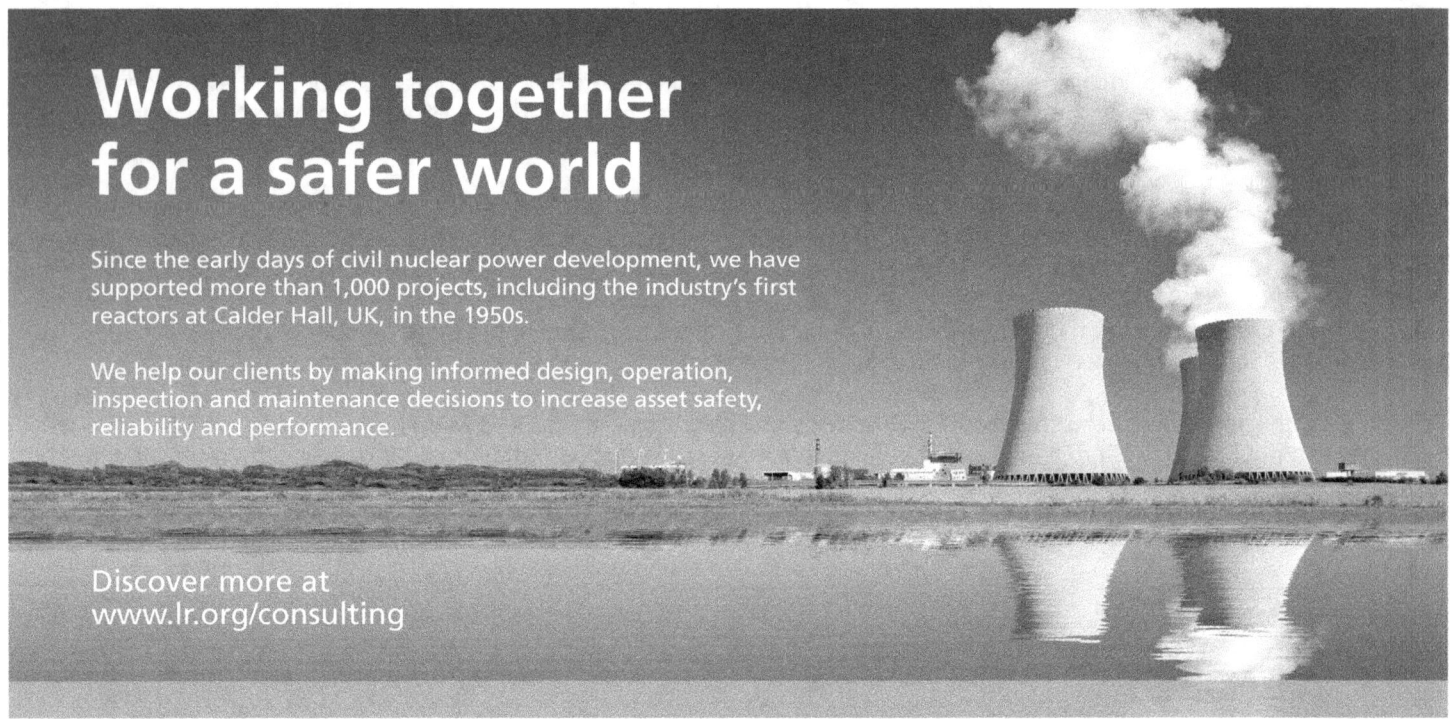

Sponsors

EPRI Assesses Seismic Resistance of Electronic Components

Together... Shaping the Future of Electricity

Technical Program Committee

Technical Program Chair: Curtis Smith, INL USA

Assistant Technical Program Chairs: Steve Epstein, Lloyd's Register Japan

                                      Vinh Dang, PSI Switzerland

                                      Ted Steinberg, QUT Australia

We would like to thank the members of the PSAM 12 Technical Program Committee. These individuals helped to make PSAM 12 a success by reviewing abstracts, technical papers, organizing sessions, and providing technical leadership for the conference.

Technical Committee Members:

Roland Akselsson	Vyacheslav S. Kharchenko
S. Massoud (Mike) Azizi	James Knudsen
Tito Bonano	Zoltan Kovacs
Ronald Boring	Ping Li
Roger Boyer	Harry Liao
Mario Brito	Francois van Loggerenberg
Kaushik Chatterjee	Jerome Lonchampt
Vinh Dang	Soliman A. Mahmoud
Claver Diallo	Diego Mandelli
Nsimah Ekanem	Donoval Mathias
Steve Epstein	Zahra Mohaghegh
Fernando Ferrante	Thor Myklebust
Federico Gabriele	Cen Nan
Ray Gallucci	Mohammad Pourgolmohammad
S. Tina Ghosh	Marina Roewekamp
David Grabaskas	Clayton Smith
Katrina Groth	Shawn St. Germain
Seth Guikema	Ted Steinberg
Steve Hess	Kurt Vedros
Christopher J. Jablonowski	Smain Yalaoui
Moosung Jae	Robert Youngblood
Jeffrey Joe	Enrico Zio

Organizing Committee

General Chair: Dr. Todd Paulos

General Vice Chair: Prof. Stephen Hora, USC

Technical Program Chair: Curtis Smith, INL USA

Webmaster, Registration,
Support for Papers/Abstracts
Submission and Review: Hanna Shapira, TICS

Table of Content

Table of Content

Human Reliability Dependency Analysis and Model Integration Process

Jan Grobbelaar[a], Michael Hirt[a], Mary Presley[b], and Chris Cragg[c]
[a] Scientech, a business unit of Curtiss-Wright Flow Control Company, Tukwila, WA, USA
[b] EPRI, Charlotte, NC, USA
[c] Cragg Consulting, Grapevine, TX, USA

Abstract: The purpose of this paper is to describe the process for integrating *HRA Calculator*[®] dependency analysis results into a *CAFTA* cutset model. Fundamental to this process is that dependencies between human failure events (HFEs) need to be addressed before cutsets are truncated to prevent inappropriate truncation of cutsets containing dependent HFEs. To prevent truncation, human error probabilities (HEPs) need to be set to high values before solving the fault tree. For a model with more than 100 post-initiator HFEs, this can be a formidable challenge due to the exponential nature of the problem – current PC hardware and software limits can be challenged. To assist in this process, the *HRA Calculator Helper* software tools for optimizing HEP values to prevent inappropriate truncation from a cutset solution and for generating cutset recovery rules files to implement dependent joint HEPs in the cutsets are discussed.

Keywords: HRA Calculator, Dependency Analysis, CAFTA, Quantification

1. BACKGROUND

Probabilistic risk assessment (PRA) is performed in the USA in accordance with the ASME/ANS PRA Standard [1], hereafter referred to as the *"PRA Standard"*, as endorsed by Regulatory Guide 1.200 [2]. All nuclear power plants are required to have PRAs and they are typically applied in risk management. EPRI has developed several software programs to build PRA models that are in widespread use in the USA and also in a number of international utilities. The suite of programs collectively known as the *Risk and Reliability (R&R) Workstation* includes software like *CAFTA* [3] and *PRAQuant* (batch processor for quantification) [4]. EPRI also developed the *HRA Calculator*[®] [5] which is a software implementation of various human reliability (HRA) methods [6, 7 and 8]. The *HRA Calculator* has a dependency analysis module that can import cutsets, identify combinations of human failure events (HFEs), apply systematic dependency rules and generate joint human error probabilities (HEPs) that need to be applied to the cutsets via post-processing as part of the PRA quantification process. This requires interfacing between the *HRA Calculator* and *CAFTA*. To facilitate this process, a software package called *HRACalculator Helper* has been developed for distribution with the *HRA Calculator*. This package contains two utilities; one for generating a *CAFTA QRecover* recovery rules file from *HRA Calculator* output, and one for generating HEP seed values used to quantify a model before applying the recovery rules file.

2. INTRODUCTION

The purpose of HRA dependency analysis as part of PRA is to identify combinations of HFEs in cutsets and to evaluate them for possible dependencies. The fundamental concern is that HFEs may not be statistically independent, which can lead to underestimation of risk metrics when using cutset methodology. The *PRA Standard* has specific high level requirements "QU-C1" to identify combinations of HFEs and "QU-C2" to evaluate them for dependencies. It is very important to note that the dependency analysis needs to be accomplished *before* cutsets are truncated. Performing a dependency analysis after cutset truncation leaves open the concern about inappropriate truncation due to joint HEPs that may be unjustifiably low. This paper describes the iterative processes developed to integrate dependency analysis results obtained from the *HRA Calculator* into a *CAFTA* model.

3. MODEL INTEGRATION PROCESS

3.1. Identification of HFE Combinations

The basic approach to avoid inappropriate truncation is to set all the post-initiator HEPs to 1.0 then re-generate the cutsets to force retention of cutsets that would normally truncate out due to low joint HEPs based on independent HEP values. If the model can solve with HEPs set to 1.0 at the same truncation level where the nominal model is convergent (less than 5% added to total risk metric at next lower order truncation level per *PRA Standard* supporting requirement "QU-B3"), all HFE combinations can be identified. However, in practice, a model with more than 100 or so HFEs may not solve with HEPs set to 1.0, given current computer hardware and software limitations. This requires both lowering the HEPs as well as truncation levels iteratively until a practical cutset solution can be obtained to start the identification process with.

The identification process is illustrated in Figure 1 below. The HEPs are initially set to 1.0 (using a flag file) in Step 1.1. The truncation level is set to a relative high value e.g. 1E-06 in Step 1.2. Generation of cutsets is then attempted at successively lower truncation levels using a *PRAQuant* batch processing file. If cutsets can be generated at the truncation level where the cutsets normally converge e.g. 1E-12, all cutsets that had previously truncated out due to low joint HEPs are retained in the solution for dependency analysis. However, if cutsets cannot be generated at this truncation level, the HEPs are decreased by 0.1 (as long as HEPs are not decreased to less than their independent values) and the next iteration is started through the cutset generation process. These iterations are continued until a cutset solution at 1E-12 is found or until the HEP value reaches 0.1. HEP values less than 0.1 are not recommended for identification purposes as too few combinations may be identified resulting in "unanalyzed" combinations appearing in the cutsets after seed values are applied, which delays convergence and/or may be unacceptable to the results, as joint HEPs for unanalyzed combinations are set to 1.0 in this approach.

In practice, the identification process can result in anything from several hundred thousand to a few million cutsets for a typical internal events PRA core damage frequency (CDF) model. The number of cutsets generated when HEPs are set to 1.0 is an exponential function of the number of HFEs in the model. Typical results for a "well behaved" CDF model with 89 HFEs are shown in Table 1, yielding 1,584,636 cutsets at 1E-12. In this case it was possible to find a solution at 1E-12 with all HEPs set to 1.0 by running a single top model. Typical results for a CDF model with 160 HFEs are shown in Table 2. In this case, a solution could not be obtained at 1E-12 with HEPs set to 1.0, which required lowering the HEPs as well as truncation level, yielding 1,317,312 cutsets at 1E-11 with HEPs at 0.1. These numbers of cutsets are handled quite adequately by *CAFTA* Version 5.4 and the *HRA Calculator* Version 5.

The cutsets are imported into the *HRA Calculator* in Step 1.3. For the examples above, 1292 HFE combinations were identified in the model with 89 HFEs with HEPs set to 1.0 and truncation level at 1E-12; while 24,331 combinations were identified in the model with 160 HFEs with HEPs set to 0.1 and truncation level at 1E-11. During this process, combinations of HFEs are programmatically identified and systematic dependency rules are applied. Various importance measures, for example risk achievement (RA), Fussel-Vesely (FV) and dependence importance (DI) are calculated [8, 9]. These importance measures are with respect to the cutset solution obtained by setting the HEPs to 1.0 (or other high values) and are useful to determine the potential impact of dependent HFE combinations should independence not apply between HFEs in a combination.

The analytical part of the HRA dependency analysis [8, 9] is performed in Step 1.4. Combinations can be ranked by the various importance measures. Combinations remain linked to the cutsets in which they are identified, so that they can be inspected by a user to perform a more detailed, contextual dependency analysis than the programmatic analysis.

**Table 1: Number of Cutsets for Identification of
HFE Combinations for Model with 89 HFEs**

Decade	By Decade	Total
1E-1 to 1E-2	2	2
1E-2 to 1E-3	4	6
1E-3 to 1E-4	8	14
1E-4 to 1E-5	55	69
1E-5 to 1E-6	232	301
1E-6 to 1E-7	1889	2190
1E-7 to 1E-8	6819	9009
1E-8 to 1E-9	31230	40239
1E-9 to 1E-10	106523	146762
1E-10 to 1E-11	340813	487575
1E-11 to 1E-12	1097061	1584636

**Table 2: Number of Cutsets for Identification of
HFE Combinations for a model with 160 HFEs**

Decade	By Decade	Total
1E-2 to 1E-3	1	1
1E-3 to 1E-4	10	11
1E-4 to 1E-5	92	103
1E-5 to 1E-6	466	569
1E-6 to 1E-7	2193	2762
1E-7 to 1E-8	12528	15290
1E-8 to 1E-9	57563	72853
1E-9 to 1E-10	246067	318920
1E-10 to 1E-11	998392	1317312

When the analytical part of the HRA dependency analysis is completed, the *HRA Calculator* is used to generate an output file with the HFE combinations and their independent and dependent joint HEPs in Step 1.5. At this point, a user can specify a minimum joint HEP to be applied. In general, the application of a minimum joint HEP is not considered necessary for *PRA Standard* Category II applications, as the application of the systematic dependency rules in the *HRA Calculator* is deemed to be sufficient to demonstrate the levels of dependence applied between dependent events. For *PRA Standard* Category I applications where no detailed dependency analysis is performed, application of a minimum joint HEP is considered necessary. This output file serves as input for the first step in the quantification process described in the next Section.

3.2 Generation of Recovery Rules File

The joint HEPs are applied to the CAFTA cutsets via a post-processing recovery rules file. The generation of the recovery rules file from HRA Calculator output is illustrated in Figure 2 and relies on the *HRACalculator Helper* tool, developed to automate this process.

The output from the *HRA Calculator* is imported into the *HRACalculator Helper* tool in Step 2.1 by user input. The user also specifies the *CAFTA* database ("RR" file) that needs to be updated by addition of the joint HFE basic events that will be applied to the cutsets as recoveries.

In Step 2.2, the recovery rules file in *CAFTA QRecover* rule file format is generated. There are two general approaches to applying the joint HEPs to the cutsets (1) replace the individual HFEs in a cutset with a single joint HFE or (2) apply a recovery factor to the cutset. The advantage of this latter approach is that the individual HFEs are retained in the cutset, which greatly facilitates cutset review and is therefore the recommended approach. There are two options for applying recovery factors; one option ("multiplier") will generate recovery rules that add a dependency factor with a value of 1 or higher, the other option will set the HEPs in an HFE combination to 1.0 and append the joint HFE event to the cutset with its joint HEP as a recovery factor.

A *CAFTA QRecover* recovery rules file that appends a multiplier recovery factor has the following attributes:

1. All HEPs that occur in HFE combinations are retained at their nominal values.
2. The number of allowed recoveries per cutsets is set to 1
3. The HFE combinations are sorted by decreasing combination order. This is necessary to ensure that the highest order combinations are recovered before lower order combinations, which may be subsets of a higher order combination. Should a lower order combination recovery factor first be applied to a higher order combination, given that only one recovery is allowed, the higher order combination would only be partially recovered.
4. The recovery factor to be applied to a combination is a multiplier which is obtained by dividing the joint dependent HEP by the joint independent HEP obtained from the *HRA Calculator* analysis.

A *CAFTA QRecover* recovery rules file that appends a joint HEP recovery factor has the following attributes:

1. All HEPs that occur in HFE combinations are set to 1.0.
2. The number of allowed recoveries per cutsets is set to 1
3. The HFE combinations are sorted by decreasing combination order. This is necessary to ensure that the highest order combinations are recovered before lower order combinations, which may be subsets of a higher order combination. Should a lower order combination recovery factor first be applied to a higher order combination, given that only one recovery is allowed, the higher order combination would only be partially recovered.
4. The recovery factor to be applied to a combination is the joint dependent HEP obtained from the *HRA Calculator* analysis.
5. Adds an independent HEP event recovery factor with the nominal value to cutsets containing only single, independent HEPs that were set to 1.0.

Following generation of the recovery rules, the joint HFE basic events are added to the *CAFTA* database in Step 2.3.

3.3 Seed Optimization

The *QRecover* recovery rules file is imported by the *HRACalculator Helper* Seed Optimizer in Step 2.4 in Figure 2. The seed optimization process was developed to reduce the HEPs - that ideally should remain at 1.0 - to lower values to improve the speed of the cutset solution, which is very desirable if the model is, for example to be used in online risk monitoring. The problem with keeping HEPs at 1.0 is that larger numbers of cutsets needs to be manipulated, which slows processing time down. However, lowering the HEP values may introduce inappropriate truncation concerns.

The philosophy behind this process is to reduce the conditional HEP for a specific HFE to a minimum value that will still ensure that the no conditional joint HEPs in the unrecovered cutsets would be less than the joint HEP that would be applied via the recovery rules file, given that all HEPs may be simultaneously reduced. If the value of the joint HEP in the unrecovered cutsets is the same as the value that would be applied via the recovery rules file, the unrecovered cutset containing the joint HEP would not be inappropriately truncated. For example, for HFEs A, B and C:

Independent HEPs: $HEP_A = 0.01$, $HEP_B = 0.0001$, $HEP_C = 0.01$
Independent Joint HEP: $HEP_A \times HEP_B \times HEP_C = 1E\text{-}08$

Conditional HEPs: $CHEP_A = 0.01$, $CHEP_B = 0.5$, $CHEP_C = 0.06$,
Conditional Joint HEP_{ABC}: $CHEP_A \times CHEP_B \times CHEP_C = 3E\text{-}04$

In the recovery rules file:

$CHEP_A = 1$, $CHEP_B = 1$, $CHEP_C = 1$
$CHEP_A \times CHEP_B \times CHEP_C \times HEP_{ABC} = 3E\text{-}04$

If HEP_A, HEP_B, and HEP_C are set to $\sqrt[3]{3E-04} = 6.7E - 02$, the product $HEP_A \times HEP_B \times HEP_C$ will be 3E-04 before truncation (instead of 1E-08), thus any unrecovered cutset that contains this combination with these values will not be truncated inappropriately. The HEP seed values for this simple example can therefore be reduced from 1.0 to 6.7E-02 without truncation concerns, and processing speed will be improved as fewer cutsets will be generated and carried through the process.

The seed values are generated in Step 2.5 of Figure 2. The seed optimization process needs to consider the impact of lowering an HEP on the joint HEP of all combinations in which the HEP occurs, in conjunction with lowering of all other HEPs. All HEPs are initially set to 1.0. A reduction factor is then recursively applied to the HEPs while all the recovery rules are tested to ensure that the joint HEP for any HFE combination is not reduced to less than the joint HEP specified in the recovery rules file. The *CAFTA* database is populated with the seed values in Step 2.6.

3.4 Quantification Process

The quantification process is illustrated in Figure 3. In Step 3.1, the initial truncation level ("Trunc") is set to the final truncation level achieved in the identification runs ("IDTrunc") from Step 1. The cutsets are generated by solving the single top model using the seed values in the database, recovery rules are applied and cutsets are truncated after applying the recovery rules in Step 3.2. If the cutsets can be generated, they are inspected for convergence by considering the change in CDF in the last decade in Step 3.6. If this is less than 5%, the cutsets obtained in the previous decade are considered converged. If the cutsets are not convergent, the truncation level is lowered and Step 3.2 is repeated.

For models with more than a 100 HFEs or so, the single top model may not solve due to current computer hardware and software limitations. The model then needs to be solved at lower logical levels and the resulting cutsets merged. A first attempt at model solution is made by solving the model on an initiating event basis in Step 3.3. If a specific initiator does not solve, an attempt is made to solve by specific event tree sequence for that initiator in Step 3.4. To reduce the number of recovery rules that need to be processed at sequence level, a sequence-specific identification run can be performed to only identify HFE combinations that are produced by the sequence, and sequence specific seed values and *QRecover* rules file can be generated. However, it is rare that one needs to solve the model at this resolution, and it might be indicative of other modeling issues that ought to be addressed to avoid this. If cutsets are generated at initiator or sequence levels, the resulting cutsets are merged and then inspected for convergence by considering the change in CDF in the last decade in Step 3.6. If this is less than 5%, the cutsets obtained in the previous decade are considered converged. If the cutsets are not convergent, the truncation level is lowered and the process is repeated.

If the cutsets are convergent in Step 3.6, they need to be checked to determine if there are any remaining "unanalyzed" combinations in Steps 3.7 and 3.8. If the truncation level ("Trunc") is equal to or higher than the truncation achieved during the HFE combination identification process ("IDTrunc), *and* if all the seed values are less than or equal to the values used in the identification process, no new HFE combinations should be produced. However, it is often the case that either the truncation level achieved during the identification process is higher than the truncation level where convergence occurs, and/or seed values are higher than the values used in the identification process. In this case, additional HFE combinations could be produced. Such combinations would have their HEPs remain at 1.0 as there would not be any recovery rules applying a joint HFE recovery factor to them. These "unanalyzed" combinations could be important if they skew the results. The efficient way to deal with them is to import (add) these cutsets to the same HRA Calculator dependency analysis database for analysis and generation of additional recovery rules by returning to Step 1.3 for another iteration through the process.

4. CONCLUSION

A systematic process has been developed for integrating HRA Calculator dependency analysis results into a *CAFTA* cutset model necessitated by computer hardware and software limitations. Although *CAFTA* was used in practice, this process could be generalized for application to any other PRA software relying on cutset methodology. The flow charts developed in this paper can serve as a basic framework for developing more detailed user guidance to accompany these software packages.

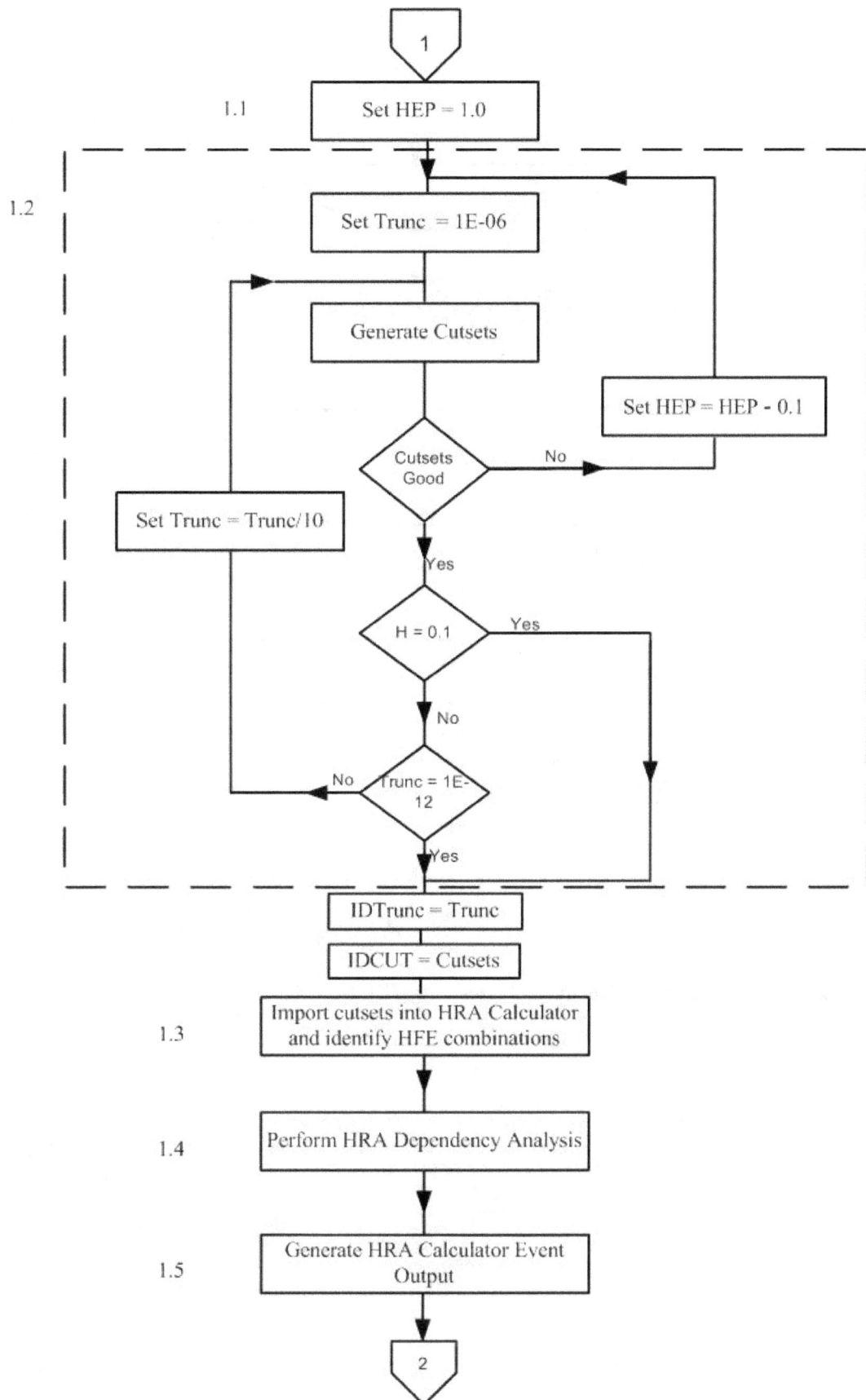

Figure 1: Identification of HFE Combinations and Dependency Analysis

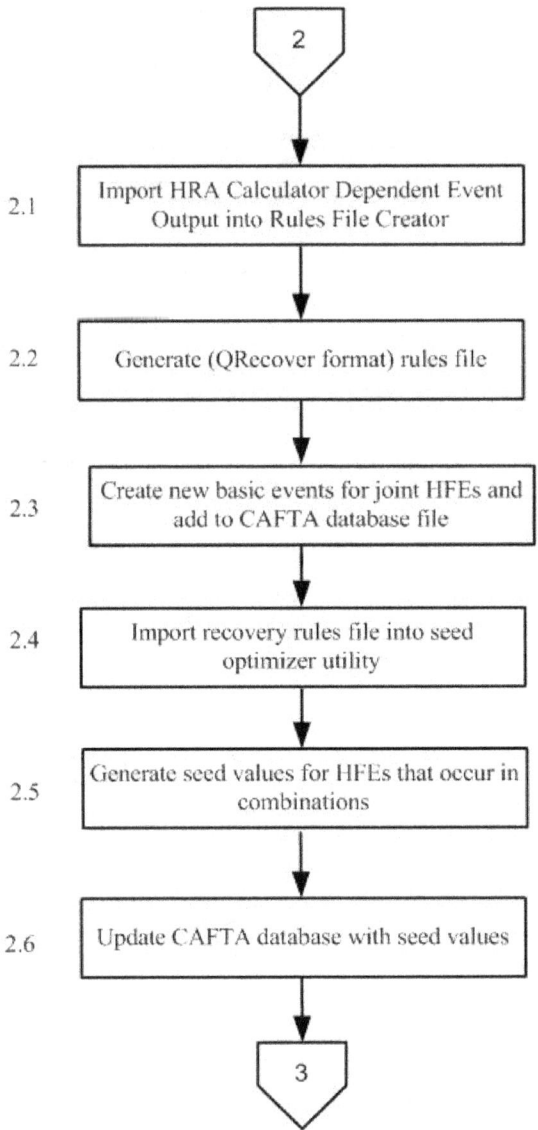

Figure 2: Recovery Rules and HEP Seed Optimization

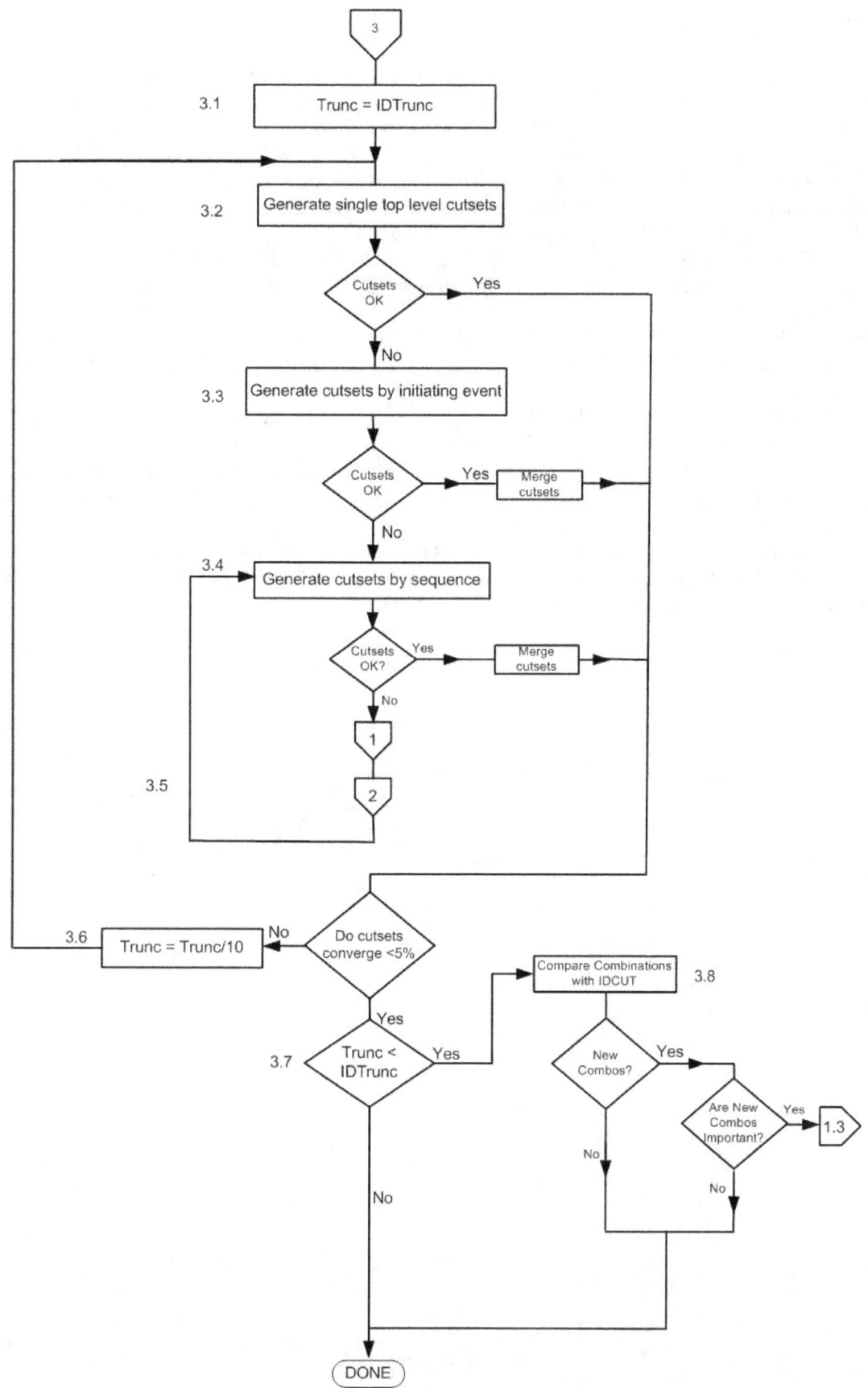

Figure 3: Quantification Process

References

[1]   ASME/ANS RA-Sa-2009, Addenda to ASME/ANS RA-S-2008, *Standard for Level 1/Large Early Release Frequency Probabilistic Risk Assessment for Nuclear Power Plant Applications*, The American Society of Mechanical Engineers, New York, NY, February 2009.

[2]   Regulatory Guide 1.200, *An Approach for Determining the Technical Adequacy of Probabilistic Risk Assessment Results for Risk-Informed Activities*, Revision 2, March 2009.

[3]   Computer Aided Fault Tree Analysis System, Version 5.4, EPRI Product ID 1009644.

[4]   PRAQuant, Version 5.01a, EPRI Product ID: 1026506

[5]   EPRI HRA Calculator, Version 5.0, EPRI Product ID: 3002000751, September 2013 TR 100259, *An Approach to the Analysis of Operator Actions in Probabilistic Risk Assessment*. EPRI, Palo Alto, CA: 1992.

[6]   NUREG/CR-1278, *Handbook of Human Reliability Analysis with Emphasis on Nuclear Power Plant Applications (THERP)*, U.S. Nuclear Regulatory Commission. A. D. Swain and H. E. Guttman, August 1983.

[7]   NUREG/CR-6883, *The SPAR-H Human Reliability Analysis Method*, U.S. Nuclear Regulatory Commission, D. Gertman, H. Blackman, J. Marble, J. Byers, C. Smith, August 2005.

[8]   J. F. Grobbelaar, J. A. Julius, F. Rahn, *Analysis of Dependent Human Failure Events Using the EPRI HRA Calculator®*, International Topical Meeting on Probabilistic Safety Assessment, PSA '05, San Francisco, California, September 11 to 15, 2005.

[9]   J. F. Grobbelaar, J. A. Julius, M. Averett, F. Rahn, *Automated Human Reliability Analysis Using the EPRI HRA Calculator®*, ANS PSA 2008 Topical Meeting – Challenges to PSA during the nuclear renaissance, Knoxville, Tennessee, September 7-11, 2008.

Formative Evaluation for Optimal Upgrades in Nuclear Power Plant Control Rooms

Ronald L. Boring[*]

Idaho National Laboratory, Idaho Falls, Idaho, USA

Abstract: As control rooms are modernized with new digital systems at nuclear power plants, it is necessary to evaluate the operator performance using these systems as part of a verification and validation process. There is no standard, predefined process available for assessing what is satisfactory operator interaction with new systems, especially during the early design stages of a new system. This paper identifies a process framework for evaluating human system interfaces as part of control room modernization. The process is geared toward generalizability to other applications and serves as a template for utilities and safety-critical industries undertaking their own control room modernization activities.

Keywords: Usability, formative, summative, control room, nuclear power plants.

1. INTRODUCTION

Main control room (MCR) modernization is a reality at nuclear power plants (NPPs). With life extensions of plants beyond the original 40-year operating licenses, there is impetus to upgrade aging systems to achieve greater efficiencies and maintain high operational reliabilities. Since existing MCRs in United States (U.S.) plants are largely analog or mechanical systems and since equivalent analog or mechanical replacements for these systems cannot be readily obtained, modernization comes in the form of digital upgrades. In particular, utilities are replacing individual analog systems on the control boards with distributed control systems (DCS) featuring digital displays, programmable logic control, and alphanumeric and touch input devices. These upgrades have to date been centered on non-safety systems, which do not require extensive license modifications through the U.S. Nuclear Regulatory Commission (NRC). Nonetheless, because the human-system interaction (HSI) between the operators and the DCS is considerably different than the analog systems it replaces, it is prudent to undertake a thorough process of ensuring the utility and performance of the new systems.

One of the key aspects influencing the effectiveness of the new DCS is the operator interaction with that system. Within the field of human factors engineering (HFE) is an area of specialization geared toward optimizing the design of the new HSI and assessing operator performance in using the new HSI. The U.S. Department of Energy (DOE) has established the Light Water Reactor Sustainability (LWRS) program to support research aimed as maintaining the current fleet of U.S. reactors through their life extension. Among the areas of research within the LWRS Program is research centered on improving instrumentation and control (I&C), including the HSI. The Control Room Modernization Pilot Project works with utilities to conduct human factors research that helps utilities determine the best I&C upgrades to their control rooms. Since the MCR is heavily dependent on operator control, control room modernization especially benefits from the operator-centered emphasis of HFE.

Previous efforts under the LWRS Control Room Modernization project have developed a generic style guide for HSI upgrades (Ulrich et al., 2012); conducted the planning and analysis activities that are essential antecedents to new design work (Hugo et al., 2013); and developed a full-scale, full-scope, reconfigurable simulator capable of being used in control room modernization studies (Boring et al., 2012 and 2013). This latter effort is particularly noteworthy, as it provides a neutral testbed that may be used by utilities to support operator studies and basic design research necessary to transition to

---

[*] Ronald.Boring@inl.gov

Figure 1: The Human System Simulation Laboratory.

digital control rooms. The resulting Human-System Simulation Laboratory (HSSL) is depicted in Figure 1 in its recently updated version. The HSSL currently supports four full plant models in a first-of-a-kind glasstop configuration that allows mimics of existing analog I&C as well as rapid development and testing of DCS technology on the virtual control panels. Individual collaborations with utilities are disseminated to ensure that HFE lessons learned benefit all interested parties, including other utilities considering control room modernization or the regulator that must review changes to control room functionality.

Previous efforts under the LWRS Control Room Modernization project have developed a generic style guide for HSI upgrades (Ulrich et al., 2012); conducted the planning and analysis activities that are essential antecedents to new design work (Hugo et al., 2013); and developed a full-scale, full-scope, reconfigurable simulator capable of being used in control room modernization studies (Boring et al., 2012 and 2013). This latter effort is particularly noteworthy, as it provides a neutral testbed that may be used by utilities to support operator studies and basic design research necessary to transition to digital control rooms. The resulting Human-System Simulation Laboratory (HSSL) is depicted in Figure 1 in its recently updated version. The HSSL currently supports four full plant models in a first-of-a-kind glasstop configuration that allows mimics of existing analog I&C as well as rapid development and testing of DCS technology on the virtual control panels. Individual collaborations with utilities are disseminated to ensure that HFE lessons learned benefit all interested parties, including other utilities considering control room modernization or the regulator that must review changes to control room functionality.

Because of the central role the operator plays in using the upgraded HSIs in the MCR, it is crucial that utilities properly design and evaluate their new systems using a vetted HFE process. However, currently available guidance on HFE for NPPs either does not address control room modernization (instead focusing on new builds) or doesn't explain how to use an iterative design-evaluate process that provides early stage feedback on a novel design. This paper highlights a staged approach, in which a series of usability evaluations are performed throughout the design life cycle of the system. This ensures the usability and ultimate utility of the control room modernization. The approach is novel to the nuclear industry, but it serves as a solid framework by which other safety-critical industries can engage in human-centered upgrades to HSIs.

2. DESIGN AND EVALUATION FOR CONTROL ROOM UPGRADES

2.1 NUREG-0711 Framework

The U.S. Nuclear Regulatory Commission (NRC) published the *Human Factors Engineering Program Review* Model in their NUREG-0711, Rev. 3 (O'Hara et al., 2012). The purpose of NUREG-0711 is to provide the procedure by which U.S. NRC staff review the effectiveness of human factors activities related to new construction and license amendments. Title 10, Parts 50 and 52, of the *Code of Federal Regulations* (10 CFR 50 and 52) provides the legal basis for requiring human factors considerations in nuclear power plant main control rooms. NUREG-0711 further defines human factors engineering as "The application of knowledge about human capabilities and limitations to designing the plant, its systems, and equipment." Put succinctly, NUREG-0711 outlines the process utilities must follow to ensure that control rooms support the activities operators need to perform.

NUREG-0711, Rev. 3, contains four general categories of activities, ranging from planning and analysis, to design, verification and validation (V&V), and implementation and operation. Each of these phases is described below:

- The *planning and analysis phase* gathers information on the system, functions, tasks, and operator actions, which help to define the requirements for the system being implemented.
- These requirements, in turn, drive the second category of activities, related to *design* of the new or modified system. The requirements are turned into a style guide and specification and are then translated into the actual HSI.
- After the system design is finalized, it must undergo *verification and validation* to ensure that the system works as designed. Importantly, from a human factors perspective, the system should also be usable by the target users of the system, which are reactor operators in the case of the MCR. V&V remains an area of confusion in the field of human factors, as the distinction between verification and validation is not always clear. Fuld (1995) suggests that verification entails confirming existing truths, while validation confirms performance. This can be understood simply to mean that verification involves checking the HSI to an existing human factors standard like NUREG-0700 (U.S. NRC, 2002), while validation requires checking the performance of the system and operators according to desired performance criteria.
- Finally, the system must be *implemented and operated*, which includes monitoring operator performance in the actual use of the system.

These four main categories of human factors activities are further subdivided into a total of 12 elements, as depicted in Table 1.

Table 1: The stages of NUREG-0711, Rev. 3.

Planning and Analysis	Design	Verification and Validation	Implementation and Operation
HFE Program Management			
Operating Experience Review			
Function Analysis & Allocation	Human-System Interface Design		
	Procedure Development	Human Factors Verification and Validation	Design Implementation
Task Analysis	Training Program Development		Human Performance Monitoring
Staffing & Qualification			
Treatment of Important Human Actions			

While NUREG-0711, Rev. 3, is an invaluable guide to the regulator as well as a roadmap for many human factors activities by the licensee, it falls short of addressing three critical areas:

1. *Types of Testing Specified:* Chapter 8 of NUREG-0711, Rev. 3, outlines the required process for human-system interface (HSI) design. The current version briefly references performing evaluations in the design phase—prior to V&V—but doesn't give detailed guidance. Specifically, Section 8.4.6 suggests there are two types of tests and evaluations that are appropriate at the design phase:

   - *Trade-off evaluations*, in which different design alternatives are considered, and
   - *Performance-based tests*, in which operator performance is assessed.

   These two are not mutually exclusive, e.g., performance-based tests can be used as part of trade-off evaluations. NUREG-0711 does not specifically require tests and evaluations during the design phase, nor does it provide examples of how such approaches are useful in shaping the design of the HSI. NUREG-0711 does require evaluation as part of the V&V activities conducted after the design phase. In particular, it advocates integrated system validation (ISV), which is "an evaluation, using performance based tests, to determine whether in integrated system's design (i.e., hardware, software, and personnel elements) meets performance requirements and supports the plant's safe operation" (O'Hara et al., 2012, p. 73). ISV is further elaborated in the earlier NUREG/CR-6393, (O'Hara et al., 1995). Note that NUREG/CR-6393 specifically states in Section 4.1.3 that the general evaluation methods used for ISV should not be used during earlier design phase activities, since they have different underlying goals. The ISV approach in NUREG-0711 and NUREG/CR-6393 has garnered criticism in terms of the limits of how well one set of test results can generalize to every possible subsequent situation (Fuld, 2007), an argument that could be extrapolated to suggest more frequent tests earlier in the process may generalize better. Still, an emerging consensus seems to be that verification works very well at the tail-end of design, while validation needs to be conducted earlier and iteratively (see, for example, Hamblin et al., 2013).

2. *Non-Safety Systems:* NUREG-0711 provides extensive guidance in Section 8.4.4.2 on control room requirements, but these requirements refer to overall systems—especially safety systems—that need to be present in the control room at design time. However, there is no guidance on individual non-safety systems. While non-safety systems (e.g., turbine control) are not subject to the same level of regulator review as safety systems (e.g., reactor control), a standardized set of good practices across both applications is desirable. There is no guidance on how to scale the approach to non-safety systems, including differences in the level of rigor expected.

3. *Modernization:* Finally, it must be noted that NUREG-0711 is optimized for reviewing initial license submittals (e.g., new builds) or license amendments (e.g., changing the operating characteristics of a required safety system). NUREG-0711 fails to provide clear guidance on modernization—replacement of an existing non-safety system—except to say that it should reasonably conform to operator expectations to minimize the need for additional training

Because guidance is missing on how to apply human factors engineering for modernization efforts on the existing fleet, the goal of this paper is to augment the guidance in NUREG-0711 specifically to address how to upgrade existing HSIs for non-safety systems as part of a NUREG-0711 compliant (albeit unrequired) process.

2.2 EPRI Guidance

The Electrical Power Research Institute (EPRI) has published useful guidance on development of a human factors engineering process in support of control room modernization. *Human Factors*

Guidance for Control Room and Digital Human-System Interface Design and Modification: Guidelines for Planning, Specification, Design, Licensing, Implementation, Training, Operation, and Maintenance, TR-1010042 (EPRI, 2005) provides thorough discussions on a number of relevant steps in modernization, including control room modernization related to hybrid control room upgrades such as featured in the current LWRS projects.

Section 3.8 of EPRI-TR-1010042 emphasizes that these activities should be performed not as a single step after the design process but as a parallel activity coinciding with design. Important steps in the assessment prior to the final ISV include:

- Section 3.8.3.1: Planning for HFE V&V
- Section 3.8.3.2: Personnel Performing HFE V&V Activities and Criteria to be Used, verification activities performed by designers and validation by independent human factors experts
- Section 3.8.3.3: HSI Inventory and Characterization (e.g., location of displays, readability of graphical elements on displays, etc.)
- Section 3.8.3.4: HSI Task support Verification, in which representative tasks to be performed on the system are tested using operators using either static or dynamic HSI display elements
- Section 3.8.3.5: HFE Design Verification of the finalized HSI against design specifications and standards
- Section 3.8.3.6: Operational Conditions Sampling, in which key aspects of personnel tasks, plant conditions, and situations as determined in the planning and analysis phase (e.g., especially from the operating experience review) are tested

Within these suggestions, ERPI-TR-1010042 provides suggestions for performance measures in Section 3.10.3.6. These include measures to catalog the actions being carried out by the operators (e.g., responding to an alarm or navigating between displays), measures of task performance (e.g., time and accuracy to complete a given task of interest), and subjective measures (e.g., operator opinions on facets of the HSI).

EPRI-TR-1010042 provides helpful additional detail not covered in NUREG-0700, tailored to the specific task of control room modernization. It also emphasizes the importance of ongoing V&V activities as part of the design process, not simply as an end-state activity to be completed after the design is finalized and implemented.

3. A SIMPLIFIED FRAMEWORK FOR EVALUATING OPERATOR PERFORMANCE

As noted, NUREG-0711 does not provide explicit guidance for conducting HSI evaluations during the design phase. Here, we outline a simplified framework to redress this shortcoming and to provide the context and methods suitable for early stage HSI evaluation in support NPP control room modernizations. The key idea featured here is that of the iterative design cycle—one in which HSIs are designed, prototyped, tested, and improved in a cyclical fashion (Nielsen, 1993). Iterative design is central to user-centered design process found in International Standards Organization (ISO) Standard 9241 that is at the heart of most human factors design activities (ISO, 2010). A core tenet of iterative design is that the resulting HSI be more usable when built through an iterative process involving early testing rather than built to completion and then tested. Feedback provided early in the design process helps to ensure that error traps in the HSI are eliminated rather than ingrained in the design, meaning it is easier to fix usability issues earlier in the design than as a fix after the design is finalized. In terms of control room modernization, the equivalent argument would be that evaluation incorporated into the design phase will produce a system more acceptable, efficient, and useful to operators rather than one that features separate design and V&V phases. The approach we advocate includes a V&V activity at the end of the design process but also incorporates small-scale V&V activities in conjunction with design milestones. Thus, V&V becomes a staged activity rather than a single terminating activity after the design.

Figure 2: An example of design phase evaluations.

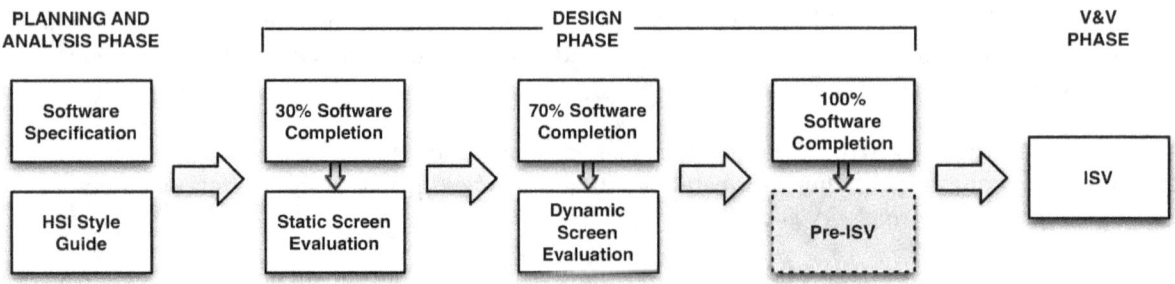

Figure 2 illustrates the idea of performing V&V activities prior to the formal ISV. In the depiction, the software specification and HSI style guide are developed based on information obtained in the planning and analysis phase. The software is then developed along three milestones during the design phase:

- At the first milestone (the 30% completion mark), the preliminary screen designs are completed. These screens can be evaluated as static screens, obtaining feedback from operators and experts on their impressions of the screen layout, look and feel, and completeness of information.
- At the second milestone (the 70% completion mark), the system dynamics are completed, and an initial functional prototype of the system may be evaluated by experts and operators. At this stage, operator performance may be assessed in use of the system.
- At the final milestone (the 100% completion mark), the system may be tested a final time (in what might be called a dry-run or pre-ISV). Or, if there is sufficient confidence in the results of the two earlier evaluations, it may be appropriate to go directly to the ISV.

There are different verification vs. validation goals for the design phase and the formal V&V phase. It is useful to think of these two phases of evaluation as formative and summative. The notion of formative vs. summative evaluation is derived from the field of education (Scriven, 1967), where it is used to distinguish between methods to assess the effectiveness of particular teaching activities (formative evaluation) vs. the overall educational outcome (summative evaluation). The approach has been widely adopted in the human factors community (Redish et al., 2002),

- *Formative Evaluation*: Refers to evaluations done during the design process with the goal of shaping and improving the design as it evolves.
- *Summative Evaluation*: Refers to evaluations done after the design process is complete with the goal of confirming the usability of the overall design.

Table 2: Verification and validation for formative and summative evaluations.

		Evaluation Phase	
		Formative	**Summative**
Evaluation Type	**Expert Review (Verification)**	Heuristic Evaluation	Design Verification
	User Testing (Validation)	Usability Testing	Integrated System Validation

ISV is, by definition, summative, and it can be concluded that the guidance in NUREG-0711 is primarily of value to summative evaluations. What, then, of formative evaluations? Table 2 outlines different verification and validation methods suitable for formative and summative evaluation. Verification is accomplished by expert review against a standard set of criteria, while validation is performed via user testing. The following considerations apply:

- *Formative Verification:* Completed during the design phase by expert review. Typical for this type of evaluation would be heuristic evaluation, which is an evaluation of the system against a pre-defined, simplified set of characteristics such as a usability checklist (Ulrich et al., 2012).
- *Summative Verification:* Completed after the design phase by expert review. Typical for this type of evaluation would be a review against applicable standards like NUREG-0700 (O'Hara et al., 2002) or requirements like the HSI style guide.
- *Formative Validation:* Completed during the design phase by user testing. Typical for this type of evaluation would be usability testing of a prototype HSI (ISO, 2010).
- *Summative Validation:* Completed after the design phase by user testing. Typical for this type of evaluation would be integrated system validation as described in NUREG-0711 (O'Hara et al., 2012).

4. Conclusions

Current guidance for HFE in support of control rooms is either focused primarily on design and evaluation for new builds or evaluation at the tail-end ISV phase. There is a need, however, to address HFE for control room upgrades and to incorporate earlier evaluation in the design cycle. By providing practical guidance on early stage design evaluation in support of control room modernization, this paper has answered two main objectives:

- To emphasize the importance of evaluation as an ongoing activity that supports design, not follows it
- To demonstrate a graded approach to HFE in which a practicable, reasonable, and cost-effective process is used to support control room modernization.

By understanding the opportunities for both verification and validation across the design life cycle of the upgrade, utilities will find a systematic and readily extensible process that ensures the success of the HSI when embarking on control room upgrades.

Disclaimer

INL is a multi-program laboratory operated by Battelle Energy Alliance LLC, for the United States Department of Energy under Contract DE-AC07-05ID14517. This information was prepared as an account of work sponsored by an agency of the U.S. Government. Neither the u.s. government nor any agency thereof, nor any of their employees, makes any warranty, expressed or implied, or assumes any legal liability or responsibility for the accuracy, completeness, or usefulness, of any information, apparatus, product, or process disclosed, or represents that its use would not infringe privately owned rights. References herein to any specific commercial product, process, or service by trade name, trade mark, manufacturer, or otherwise, does not necessarily constitute or imply its endorsement, recommendation, or favoring by the U.S. Government or any agency thereof. The views and opinions of authors expressed herein do not necessarily state or reflect those of the U.S. government or any agency thereof.

References

Boring, R.L., Agarwal, V., Joe, J.C., and Persensky, J.J. (2012). *Digital Full-Scope Mockup of a Conventional Nuclear Power Plant Control Room, Phase 1: Installation of a Utility Simulator at the Idaho National Laboratory, INL/EXT-12-26367.* Idaho Falls: Idaho National Laboratory.

Boring, R., Agarwal, V., Fitzgerald, K., Hugo, J., and Hallbert, B. (2013). *Digital Full-Scope Simulation of a Conventional Nuclear Power Plant Control Room, Phase 2: Installation of a Reconfigurable Simulator to Support Nuclear Plant Sustainability, INL/EXT-13-28432.* Idaho Falls: Idaho National Laboratory.

Electrical Power Research Institute. (2005). *Human Factors Guidance for Control Room and Digital Human-System Interface Design and Modification, 1010042.* Palo Alto: Electrical Power Research Institute.

Fuld, R.B. (2007). On system validity, quasi-experiments, and safety: A critique of NUREG/CR-6393 *International Journal of Risk Assessment and Management, 7,* 367-381.

International Standards Organization. (2010). *Ergonomics of Human-System Interaction—Part 210: Human Centred Design for Interactive Systems, ISO 9241-210.* Geneva: International Standards Organization.

O'Hara, J.M., Higgins, J.C., Fleger, S.A., and Pieringer, P.A. (2012). *Human Factors Engineering Program Review Model, NUREG-0711, Rev. 3.* Washington, DC: U.S. Nuclear Regulatory Commission.

O'Hara, J.M., Brown, W.S., Lewis, P.M., and Persensky, J.J. (2002). *Human-System Interface Design Review Guidelines, NUREG-0700, Rev. 2.* Washington, DC: U.S. Nuclear Regulatory Commission.

O'Hara, J., Stubler, W., Higgins, J., and Brown, W. (1995). *Integrated System Validation: Methodology and Review Criteria, NUREG/CR-6393.* Washington, DC: U.S. Nuclear Regulatory Commission.

Redish, J., Bias, R.G., Bailey, R., Molich, R., Dumas, J., and Spool, J.M. (2002). Usability in practice: Formative usability evaluations—Evolution and revolution. *Proceedings of the Human Factors in Computing Systems Conference (CHI 2002),* 885-890.

Scriven, Michael (1967). The methodology of evaluation. In Stake, R. E. (ed.), *Curriculum Evaluation.* Chicago: Rand McNally.

Ulrich, T., Boring, R., Phoenix, W., DeHority, E., Whiting, T., Morrell, J., and Backstrom, R. (2012). *Applying Human Factors Evaluation and Design Guidance to a Nuclear Power Plant Digital Control System, INL/EXT-12-26787.* Idaho Falls: Idaho National Laboratory.

Research on HRA methods and application for digital human-system interfaces design

Xiufeng Tian, Xingwei Jiang, Jinggong Liu
CNNC, China Nuclear Power Engineering Co., Ltd
No.117 Xisanhuanbeilu, Haidian District, Beijing, P.R.China, 100840.
tianxf@cnpe.cc, jiangxw@cnpe.cc, liujga@cnpe.cc

ABSTRACT

Operators of nuclear power plant (NPP) play a vital role in the productive, efficient, and safe generation of electric power. More widespread use of digital technology is expected in the nuclear plants, especially main control rooms (MCR). Operators face a significant challenge in digital control rooms that will be produced at various stages of instrumentation and control modernization. It is believed that the introduction of digital I&C can lead to an overall improvement in operator performance and reduce workload in abnormal conditions. However some negative consequences will also arise due to faulty HSI design based on our research and other published research.

Human reliability analysis (HRA) is a technique to evaluate the reliability of the human actions, including those actions taken by the operators in the main control room. HRA can seek to evaluate the potential for, and mechanisms of, human error that may affect plant safety. Thus, it is an essential element in achieving the Human factors engineering (HFE) design goal of providing HSI that will minimize personnel errors, allow their detection, and provide recovery capability.

The paper discusses the findings of an investigation to operating and as-building plants in China installed with fully digital I&C systems. Interviews were made with the simulator instructors, control room operators, designers of Main Control Room (MCR) about the control layout, computer interface, alarms, and procedures to understand the effects on operator performance. Specific performance shaping factors (PSFs) for digital I&C control room are proposed to be considered in HRA methods. It is also suggested how to apply the specific PSFs in digital HFE/HSI design process.

Key Words: NPP, Human factors engineering, HFE, Human reliability analysis, HRA

1 INTRODUCTION

Nowadays more widespread use of digital technology exists in the operation and design nuclear power plants. Digital human-system interfaces (HSIs) are also introduced into NPPs. Advanced main control rooms (MCR) is substituted for conventional MCR as the vital parts of a nuclear power plant with which personnel interact in performing their functions and tasks. Operators face a significant challenge in digital control rooms. A study [1] about digital and conventional HSIs by NRC indicated that the new HSIs provided positive support for crew performance, reduced workload, and were well accepted by the crews. While another finding of the study is one of the more significant effects introduced by the advanced HSI systems was on crew structure and communication. These changes of crew structure and communication have potential implications for human performance and reliability. A research [2] by BNL about Computer-Based Systems found evidence of two forms of negative effects: (1) primary task (which refers to process monitoring and control) performance declines because operator attention is directed toward the interface management task, and (2) under high workload, operators minimize their performance of interface management tasks, thus failing to retrieve potentially important information for their primary tasks. Further, these effects were found to have potential negative effect on safety. More researches have demonstrated that many uncertainties about human performance can be induced by the use of the digital HSIs.

To reduce the negative effect digital HSIs for human performance, good human performance goals should be satisfied. Some guidelines were submitted to help operators and suppliers plan, specify, design, implement, operate, maintain, and train for the modernization of control rooms and other HSI in a way that takes advantage of digital system and HSI technologies, and addresses issues concerning digital HSIs, for example NUREG/CR-0711 among which HRA is one of the 12 elements.

HRA can be used as an evaluation tool to identify vulnerabilities to human error or human engineering deficiencies of the HSIs. HRA for the new MCR should be able to consider the possible effects of new HSIs on the operator performances. But few studies are conducted so far in the HRA domain to reflect the operator performance under the digital HSIs. Most currently available human error data are collected in the operations of the current plants and simulators. The most widely used human error probabilities (HEPs) in HRA community are those in THERP handbook (NUREG/CR-1278, 1983), in which the data are collected 20 years ago without any information about the human performance dealing with the digital systems. It is necessary to study the characteristics of human performance in digital HSIs to get more information about when, where and how operators will fail and what is the risk contribution associated with these human actions.

The objective of this paper is to characterize the salient features of the digital HSIs, understand the effects on operator performance, raise specific performance shaping factors (PSFs) in HRA methods for the digital HSIs, and give a proposal to apply the specific PSFs in digital HFE/HSI design process.

2 THE CHARACTERISTICS OF DIGITAL HSIS

Currently, nuclear power plants in many countries are rapidly developing digital technology and digital HSIs is being applied in their control rooms. A survey of operating and constructing digital nuclear power plants in China indicates that common characteristics exist in digital HSIs of different reactors using different digital I&C systems.

The digital HSIs which satisfy the HFE principles in NUREG-0700, incorporate features such as soft controls, information display, computer-based procedures, computer-based alarms, touch-screen interfaces, sit-down computer workstations, and large-screen overview displays.

Table I summarizes the general characteristics of a well-designed HSIs.

To evaluate the impact of the digital HSIs on human performance and plant safety, the characteristics of the digital HSIs are described from graphic-based information display system, computer-based alarm system, and computer-based procedure system, which are necessary when operators implement required tasks.

2.1 Graphic-based Display System

Digital HSIs provide information of the process in form of readily and quickly understandable graphics-based display system on visual display units (VDUs) of workstations and on large screens. It annunciates incipient faults in the plant process and provides support for fast fault rectification.

Operators monitor the plant through screen-based displays selected from networks of hundreds or even thousands of display pages. Control of plant equipment is accomplished through soft controls that can be accessed through computer workstations.

The displays and sheets on the workstation are state-of-the-art graphic monitors using the windowing philosophy. The data are displayed in such a way that the operator can at a glance see the state of the data being displayed (e.g. back-up value, invalid value, operating state, etc.).

The display system supports each operator to choose the functions with which she/he is the most comfortable. The following are examples of the display formats available:

1. Overviews displays
2. Plant and Process displays,
3. Status displays,
4. Operator aid displays/sheets,
5. Dynamic logic and sequence Diagrams,
6. Trends, etc.

The display system follows the HFE principles of the display layout and organization in NUREG-0700.

2.2 Computer-based Alarm System

Digital HSIs contains computer-based alarm systems used to analyze, process, and reduce alarms. This requires HSI facilities to interface with alarms systems to sort alarms, view suppressed alarms, query alarm logic, modify set-points, and establish temporary alarms.

The alarm function allows fast recognition of importance in terms of necessary operator response. Classification of alarms allows for a fast recognition of their safety importance and alarm areas provide for grouping alarms according to their process system.

Alarms are indicated on the workstation screens in Alarm Sequence Display, Common Alarm Indication in the header of displays, and Operating displays, accompanied by an acoustic signal.

2.3 Computer-based Procedure System

In the operating plant investigated, operators still use paper procedures, but procedures in other constructing plants have been computerized in China.

Computer-based procedure systems provide access and display plant data referenced by procedure steps and resolve the logic of individual steps.

Computer-based procedures are likely to allow control actions are taken directly from the procedure display, or they may be semi-automated, with the operator authorizing the procedure's embedded control functions to take actions.

3 HRA CONSIDERATIONS AND PERFORMANCE SHAPPING FACTORS IN DIGITAL MCR

Fig. 1 reveals the interrelation of HSIs' characteristics and Human Response Model, and improvement of HRA application is focus on alarms, displays and procedures.

3.1 Discussion of the HRA Methods

HRA is performed as of NPP PSA to produce probability estimates for human error events (HEPs). In determining HEPs, most HRA methods account for the contextual aspects that contribute to human failure through the acknowledgment of plant conditions and performance-shaping factors (PSFs) potentially present during a task execution. The process used in most HRA methods to estimate an HEP for a task of interest is to first, estimate the base HEP (referred to as a nominal or conditional probability by some methods); second, to define the set of PSFs that affect that task; and third, to identify the significance (i.e. the size of the effect) of each PSF and to combine the effects of these PSFs to modify the base HEP for that task. Such a procedure is employed in many conventional HRA methods, e.g. THERP, ASEP, HCR, CREAM, HEART, CAHR and SPAR-H.

Most conventional HRA methods provided analysis data acquired by operating the current plants and simulators. It means that those methods do not pay attentions to the influence of digital HSIs. As a result, conventional HRA methods have limitations in considering the possibility of operators' unsafe actions due to digital HSIs and integrating with HFE activities in digital HSIs design.

To incorporate the interdependency of digital I&C systems and human operators, we believe that the current HRA methods should be modified and more applicable quantitative models for the human error assessment in digital HSIs, are necessary.

3.2 PSFs

As study results, new PSFs are recommended to improve in conventional HRA methods. The new PSFs are the three following:

PSF_D - used to evaluate display systems;

PSF_A - used to evaluate alarm systems;

PSF_P - used to evaluate computer-based procedures systems.

According to the principles in table II, new PSFs can be quantified and applied to evaluate HEP in digital HSIs.

Table I. General characteristics of well-designed HSIs and the relevant Human Response Model

	General Characteristics of a Well-Designed HSI	Human Response Model
Characteristics	**Description**	
• Accurately represents the plant	To be consistent with and supports a user's understanding and awareness of the system, its status, and the relationship between individual system elements	• Detection 　To realize an abnormal scenario occur based on alert or unpredicted information display
• Meets user expectations	To accord with HFE principles and fully enhance the work efficiency	• Diagnosis and Decision-Making 　Using computerized HSIs, in support of computerized procedures, to make the diagnosis detection to ascertain the actual plant scenarios and the necessary response for next step
• Supports situation awareness and crew task performance	Fully support users to accomplish their primary tasks of monitoring, situation assessment, response planning and response execution by providing alerts, information, procedural guidance, and controls when and where they are needed	
• Minimizes secondary tasks and distractions	Users should not need to shift attention from their primary tasks to the interface. Therefore, the need for users to perform secondary tasks such as window manipulation, display selection, and navigation should be minimized as much as possible	• Perform detail actions 　To perform certain measurements or series actions to eliminate system fault or alleviate the sequent of abnormal scenario to ensure the plant safety
• Balances workload	Optimize function allocation between human and machine to maximum enhance the human-machine efficiency	
• Is compatible with users' cognitive and physical characteristics	To accommodate human physiological and cognitive characteristics and limitations such as visual/auditory perception and anthropometrics and biomechanics	
• Provides tolerance to error	To minimize the occurrence of user errors and provides a way for users to detect and correct errors when they do occur	
• Provides simplicity	Simplest design to meet the task requirements and potentially distracting features such as excessive decorative detail or nonfunctional icons should be avoided	
• Provides standardization and consistency	Standardization and consistency make the HSI predictable and predictability lowers the workload associated with using the interface, leaving more attention for doing the primary tasks	
• Provides timeliness	To ensure that tasks can be performed within the time required and this requires consideration of the user's capabilities and system-related time constraints	
• Provides openness and feedback	Help users easily understand and track the plant process	
• Provides guidance and support	Provide an effective "help" function on line or off line to help users understand and interact with the HSI	
• Provides appropriate HSI flexibility	Computer-based HSIs can be tailored to better meet the demands of the user's ongoing tasks and to accommodate personal preferences	

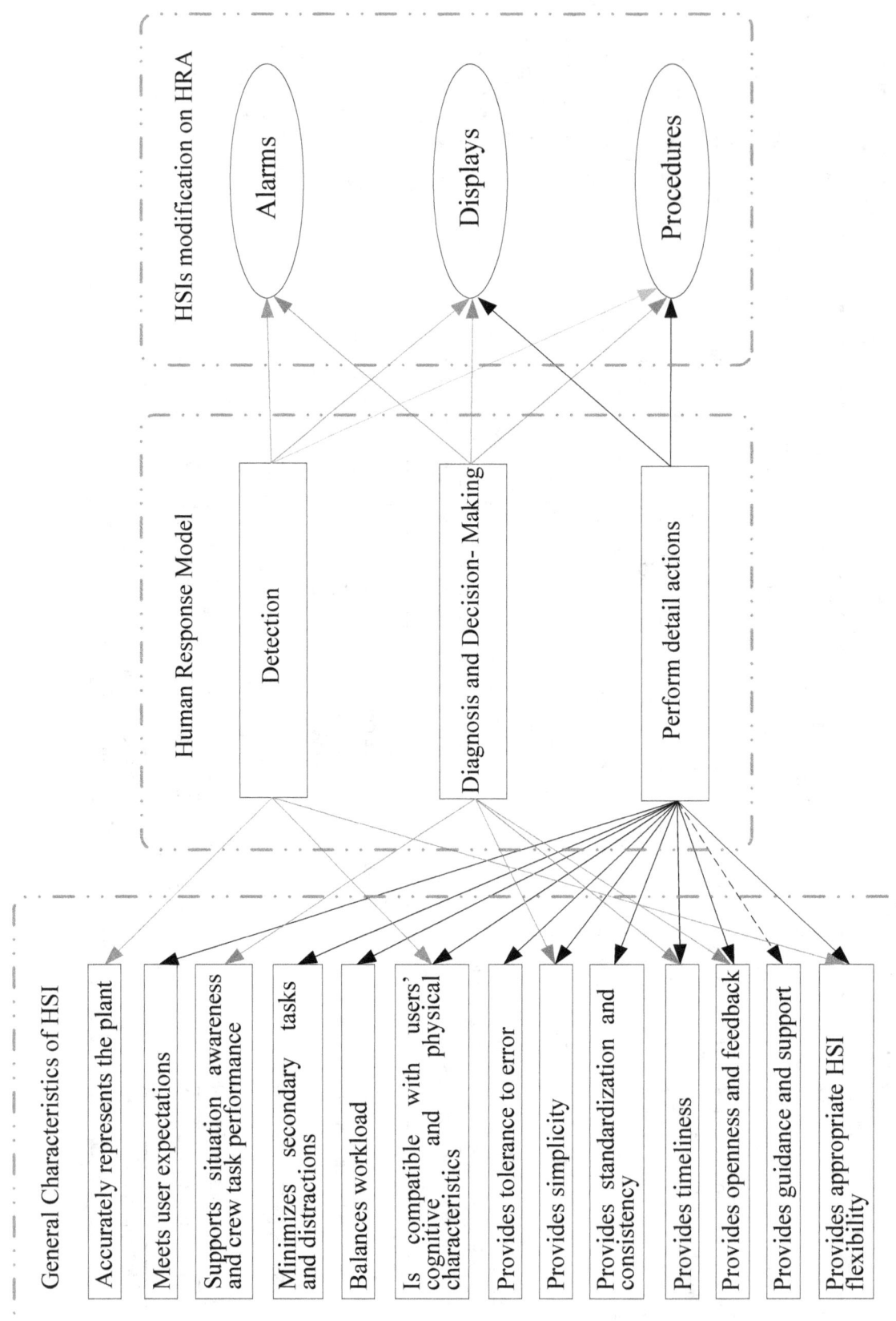

Figure 1. Interrelation of HSIs' characteristics and the Human Response Model

Table II. The quantification principles of new PSFs

PSF	Optimum Conditions	The Quantification of PSF
PSF_D	1. Users can quickly turn into the right display by 3 times mouse Clicks or less. 2. Display formats and elements will not influence the occurrence of visual fatigue. 3. Display packing density should not exceed 50 percent. Display arrangement is clear, and displayed information provides only necessary and immediately usable data. Thus users can quickly operate right equipment. 4. High-level displays can be applied to improve accuracy and efficiency.	When the evaluated displays satisfy 3 optimum conditions at least, $PSF_D=0.5$; When the evaluated displays satisfy 2 optimum conditions, $PSF_D=1$; When the evaluated displays satisfy less than 2 optimum conditions, $PSF_D=2$;
PSF_A	1. Alarms classified and optimized in reason make users easy to identify the significant alarms and respond quickly when the several alarms appear at the same time. 2. Importance of alarms is distinguished by color, voice, or description , so that users can first deal with the most important alarms on safety operation. 3. Alarms are independent and every alarm definition is clarity, thus users can fast affirm and correctly respond alarms. 4. Users can rapidly get to computer-based procedures via their direct links.	When the evaluated computer-based alarms satisfy 3 optimum conditions, $PSF_D=0.5$; When the evaluated computer-based alarms satisfy 2 optimum conditions, $PSF_D=1$; When the evaluated computer-based alarms satisfy less than 2 optimum conditions, $PSF_D=2$;
PSF_P	1. Users can rapidly get to computer-based procedures by 3 mouse clicks. 2. Computer-based procedures can be implemented efficiently and accurately by providing information displays which contain concise steps, the warnings and cautions, embedded real-time Data .etc. 3. Procedure steps can be automatically executed by system, thus avoid errors of human action. 4. Procedure steps are easy to be tracked by users, thus avoid errors of omitting steps.	When the evaluated Computer-based procedures satisfy 3 optimum conditions at least, $PSF_D=0.5$; When the evaluated Computer-based procedures satisfy 2 optimum conditions, $PSF_D=1$; When the evaluated Computer-based procedures satisfy less than 2 optimum conditions, $PSF_D=2$;

4   THE APPLICATION OF NEW PSFS

4.1 Improvement of HRA Methods

To simplify description, PSF_{HSI} ($PSF_{HSI}=PSF_D \times PSF_A \times PSF_P$) will be introduced to modify convention -al HRA methods such as HCR, THERP and SPAR-H.

For HCR model, analysts can affirm the coefficient of K_3 (a PSF related quality of operator/plant interface) by the value of PSF_{HSI}. When PSF_{HSI} is less than 1, the coefficient of K_3 is -0.22 (Excellent); when PSF_{HSI} is equal to 1, the coefficient of K_3 is 0 (Good); when PSF_{HSI} is greater than 1, the coefficient of K_3 is 0.44 (Fair);

For improvement of THERP, the BHEP should be modified by the PSF mentioned in THERP book and PSF_{HSI}.

For SPAR-H, PSF_{HSI} can be applied to affirm PSF level of ergonomics/HMI the same as HCR model.

4.2 Integration of HRA with Design

It is known that errors resulting from human factors deficiencies such as poor control room design, procedure, and training are an important contributing factor to NPP incidents and accidents. Therefore a good HFE in a NPP is an essential part to ensure public health and safety.

The HFE Program Review Model (NUREG-0711, 2004) consists of twelve review elements which provide detailed review criteria of HFE. HRA is required as one of 12th elements.

To integrate the HRA results with the HFE design of digital HSIs, the HRA should be conducted through a systematic process in identification of the performance shaping factors, task analyses, quantification method, and dependence analysis. The identified critical actions and risk important actions in HRA/PSA process should have enough detailed information to support HFE amendments through human system interface design, procedure development and training. After the HFE design review is completed, there is a need to re-evaluate and possibly re-quantify the HRA/PSA [3].

HRA methods improved by New PSFs are applied HFE activities of digital HSIs. It has much more pertinence and maneuverability to modify digital HSIs design.

5   CONCLUSIONS

The digital HSIs applied in NPPs offer potential for improved operator performance, however if not appropriately applied, they may introduce new burdens for the operator. Existing HRA methods are limited to evaluate the influence of digital HSIs on operator performance, and are difficult to give out advisable suggestion tending to the improvement of digital HSIs.

In this paper, new PSFs are introduced into conventional HRA methods and used for the evaluation of human performance in digital HSIs. The HCR and THERP improved are applied to HRA in several NPP design projects in China.

More uncertainties about human performance can be induced by the wide use of the digital techniques, which lack enough practical experiences.

The improvement of HRA methods cannot evaluate all the change of human performance in digital HSIs. More real and reasonable HRA models are expected in future.

6 REFERENCES

1. E. Roth, and J. M. O'Hara, *Integrating Digital and Conventional Human-System Interfaces: Lessons Learned from a Control Room Modernization Program* (NUREG/CR-6749), Washington, DC (2001).

2. J. M. O'Hara, W. S. Brown, P. M. Lewis, and J. J. Persensky, *The Effects of Interface Management Tasks on Crew Performance and Safety in Complex, Computer-Based Systems: Overview and Main Findings* (NUREG/CR-6690, BNL-NUREG-52656), Washington, DC (2002).

3. X. He & J. Tong. *Some Considerations for Implementing Human Reliability Analysis for Advanced reactors*. ESERL (2006).

4. NRC. *Human Factors Engineering Program Review Model* (NUREG 0711), Washington, DC (2004).

5. NRC. *Human-System Interface Design Review Guidelines* (NUREG 0700), Washington, DC (2002).

6. A. D. Swain, H. E. Guttmann. *Handbook of Human Reliability Analysis with Emphasis on Nuclear Power Plant Applications* (NUREG/CR-1278), Washington, DC (1983).

A Methodology for Safety Culture Index Assessment Using Minimal Cut Sets

Kiyoon Han, Yongjin Lee and Moosung Jae*

Department of Nuclear Engineering, Hanyang University, Seoul, 133-791, Korea
*Corresponding author: jae@hanyang.ac.kr

Abstract: The purpose of this study is to evaluate the Safety Culture Impact Index (SCII) for several types of nuclear power plants in Korea. The SCII model can be used for measuring the changes of the core damage frequency which might be affected by the status of safety culture in nuclear power plants. In order to develop the SCII model, the safety culture indicators and their assessing method are developed and applied to a reference plant. The reference plants are selected and their basic events are evaluated according to the level of the impact of safety culture. The results include the procedure to obtain the safety culture impact index as well as the frequencies of accident sequences which are expressed by the logical combination of minimal cut sets. The SAREX code is used for producing safety culture impact index related to the plant status. The correlation between the basic events caused by the quality of safety culture has been analyzed in this study. The uncertainty in safety culture impact has been also analyzed. The developed SCII model might contribute to comparing the level of the safety culture among nuclear power plants as well as to improving the management safety of nuclear power plants.

Keywords: Safety Culture, Human Errors, Minimal Cut Sets, Risk, Impact Index, Nuclear Power Plants

1. INTRODUCTION

Safety culture is defined to be fundamental attitudes and behaviours of the plant staff which demonstrate that nuclear safety is the most important consideration in all activities conducted in nuclear power operation. Recently, the safety culture of nuclear power plant has been emphasized in reactor safety world-widely. Moreover, through several accidents of nuclear power plant including the Fukusima Daiichi in 2011 and Chernovyl accidents in 1986, the safety of nuclear power plant is emerging into a matter of interest. From the accident review report, it can be easily found out that safety culture is important and one of dominant contributors to accidents. It is also known that the enforcement of safety culture have an important role for improving the safety of nuclear power plant.

The term "safety culture" was first introduced by International Nuclear Safety Advisory Group (INSAG) that consists of international experts to analysis and to prevent nuclear accidents. The safety culture was defined by the INSAG as "the assembly of characteristics and attitudes in organizations and individuals which establishes that, as an overriding priority, nuclear plant safety issues receive the attention warranted by their significance" [1]. The safety culture assessment has been usually conducted using the questionnaire and the interview which are such as ASCOT and SCART. These methods by the way have some disadvantage that the subjective judgment plays an important role in safety culture assessment. The various quantitative methods for assessing safety culture are suggested in several research papers to improve this disadvantage. One of the previous research works in these areas includes the work process analysis model which evaluates the impact of organizational factors on risk using Probabilistic Safety Assessment [2]. The success likelihood index method used in the human reliability analysis (HRA) is utilized in this WPAM method. When the success likelihood index method is also a subjective oriented method in which the probability of component failure and initiating frequencies might be non-systematic and overestimated. Therefore, the purpose of this study is to develop a new methodology that assesses quantitatively the safety culture impact index overcoming these disadvantages.

2. METHODOLOGY

2.1. Definition of safety culture indicator

To achieve the main objective of this study, the methodology to produce the safety culture indicators are developed in the beginning. The safety culture indicators that show the status of safety culture in nuclear power plants are presented in various forms in the literatures [1]. INSAG-4, "Safety Culture" describes safety culture elements classifying in three categories: individual's commitment, manager's commitment, and policy level commitment. In addition, the safety culture indicators are explained to encourage self-examination in individuals and organizations [2]. Their indicators are provided as typically "yes / no" question format.

Institute of Nuclear Power Operations published "Principles of a Strong Nuclear Safety Culture" in 2004. In this reference, the definitions of eight safety culture principles and their attributes to assess the level of safety culture are specified [3]. INPO another publication "Traits of a Healthy Nuclear Safety Culture" describes the essential traits and attributes of a healthy nuclear safety culture. The traits described in that reference are divided into three categories that are similar to the three categories of safety culture in INSAG-4, "Safety Culture". The categories and their primary traits are as follows: Individual commitment to Safety, Management Commitment to Safety, Management Systems. Traits are defined as a pattern of thinking, feeling, and behaving such that safety is emphasized over competing priorities. Personal and organizational traits described in Ref. [4] are present in a positive safety culture and that shortfalls in these traits and attributes contribute significantly to the occurrence of the plant incidents.

In 2005, the Nuclear Regulatory Commission conducted a public meeting on the agency's initiatives to enhance the Reactor Oversight Process to more fully address safety culture. The USNRC staff asked stakeholders to provide suggestions/comments on the draft Safety Culture Attributes Table on a feedback form located on the Safety Culture web page. Safety Culture Attributes Table is composed of four attributes and each of them has several factors such as elementary safety culture, potential safety culture inspection information and potential safety culture measure [5].

Recently Korea Institute of Nuclear Safety developed the safety culture assessment methodology that has six indicators and thirty evaluation items [6]. The feature of this methodology utilizes the objective data: the number of safety culture self-assessment, the number of staff, the training time etc. The results produced by KINS consist of the attributes, the traits, and indicators to evaluate the safety culture of the plant organization. In this study, the safety culture indicators are developed and applied to the reference plant. The level of and traits of a Healthy Nuclear Safety Culture are surveyed and safety culture indictors and their definitions are presented in Table 1.

Table 1: Safety culture indicators and definitions

Category	Safety Culture Indicator	Definition
Individual Commitment to Safety	Human error	Prevention of human error
	Communication	Efficiency of exchanging information
	Attitude	Behaviour toward nuclear safety
Management Commitment to Safety	Highlighting safety	Operation that keeps safety as the overriding priority
	Resource	Magnitude of the human resource
Management System	Training	Degree of training for safe operation
	Procedure	Propriety of procedure to prevent unexpected accident
	Man Machine Interface	Interface level that helps staff to use machines easily

Table 2 shows the comparison between the current study of safety culture indicators and those of other international studies. There is only one study considered for human error affected by the safety culture. Mostly there is no sincere consideration for the man machine interface. However, in this study, they are considered and modelled because of the dominant importance in the nuclear safety culture.

Table 2: Comparison among safety culture indicators considered by various research organizations

Category	Safety Culture Indicator	INPO	IAEA	NRC	KINS
Individual Commitment to Safety	Human error	-	-	√	-
	Communication	√	√	-	√
	Attitude	√	√	√	-
Management Commitment to Safety	Highlighting safety	√	√	√	√
	Resource	√	√	√	√
Management System	Training	√	√	√	√
	Procedure	√	√	√	√
	Man Machine Interface	-	-	-	-

2.2. SCI assessment

The data related to the evaluation of the safety culture indicators and human errors occurring in nuclear power plants are obtained from Korea Institute of Nuclear Safety research report. The Korea Institute of Nuclear Safety which is a nuclear regulatory agency evaluates the nuclear safety in detail through the periodic inspection. They also used to present recommendations to licensee by evaluating the causes and the reasons when the reactor stops unexpectedly. The nuclear power plant assessment for the current status has been openly published through the website and it contributes to being valuable information about the current plant safety.

The methodology to evaluate the human errors entitled to "A Standard Method for Human Reliability Analysis of Nuclear Power Plants" developed in KAERI is now utilized in performing PSA in Korea [7]. This methodology explains in detail the performance shaping factor for each human errors. It presents their rating criteria. In addition, it gives information that is the relative rating of performance shaping factors analysed by the human error experts. The data and the HRA results obtained by the periodic inspection are used to develop the quantitative safety culture assessment methodology as shown in Table 3 below.

Table 3: Safety culture indicator assessment methodology

Safety Culture Indicator	Assessment Method	Descriptions
Human error	$\left(1 - \dfrac{X}{Y}\right) \times 10$	X : the number of unexpected shutdown caused by human error Y : the number of unexpected shutdown
Communication	$\left(1 - \dfrac{X}{Y}\right) \times 10$	X : the number of comments and recommendation about "communication" in periodic inspection report Y : the number of periodic inspection report (whole plant)
Attitude	$\left(1 - \dfrac{X}{Y}\right) \times 10$	X : the number of passive shutdown in unexpected situation Y : the number of unexpected shutdown
Highlighting safety	$\left(1 - \dfrac{X}{Y}\right) \times 10$	X : the number of unexpected shutdown above INES level 0 Y : the number of unexpected shutdown
Resource	$\left(1 - \dfrac{X}{Y}\right) \times 10$	X : the number of staff Y : the maximum number of staff
Training	$\left(1 - \dfrac{X}{Y}\right) \times 5 + Z$	X : the number of comments and recommendation about "training" in periodic inspection report Y : the number of periodic inspection report (whole plant) Z : "training score" from human reliability report
Procedure	$\left(1 - \dfrac{X}{Y}\right) \times 5 + Z$	X : the number of comments and recommendation about "procedure" in periodic inspection report Y : the number of periodic inspection report (whole plant) Z : "procedure score" from human reliability report
Man Machine Interface	$\left(1 - \dfrac{X}{Y}\right) \times 5 + Z$	X : the number of comments and recommendation about "man machine interface" in periodic inspection report Y : the number of periodic inspection report (whole plant) Z : "man machine interface score" from human reliability report

The data of the variable, "Z" can be obtained from the conversion of the performance shaping factor rating to the score. The performance shaping factor has a rating of three steps such as high, middle and low. The rating level of "high, middle, and low" cases has been converted to a score "5, 3, 1", respectively. The human error events obtained from the reference of human reliability analysis cited in Ref. 7 have been analysed to have each score in which the average value denotes the data "Z".

2.3. Safety Culture Impact Index model

The core damage frequency which is one of important results of the Probabilistic Safety Assessment is used for quantifying the safety culture impact in this study. The CDF which is one of important measures is obtained from the accident sequence analysis. The main process to get the CDF is to identify and quantify the minimal cut sets which are composed of a lot of basic events. To achieve this process, basic events composing the minimal cut sets are assumed to be independent. However, this assumption is not true because there should be the correlation between those basic events. The occurrences of two failure events are not independent, for example. They have correlation if they are under operation in the same temperature or pressure conditions and environments. In that case, the temperature or pressure can be a common factor between two components. Likewise, the concept of safety culture can have a common factor between human errors and component failures. That means there are correlations between basic events that have the common factor in safety culture elements. The common uncertainty source method is utilized to consider these correlation caused by the complicated safety culture [8]. The basic event used in this study is a lognormal distribution for the uncertainty analysis. This method calculates the minimal cut sets incorporating the correlation between the lognormal distributions. It is judged to be appropriate method because it can be applied to assessing the impact of the safety culture in nuclear power plants. The formula used in this study is as follows.

$$X_i = m_i X_{i0} \sum_{j=1}^{n} X_{\cdot j}^{\sigma_{ij}/\sigma_{\cdot j}} \tag{1}$$

$$\rho_{ij} = \sigma_{ij}^2 / \sigma_i^2 \tag{2}$$

$$\sigma_{ij} = \sigma_i \sqrt{\rho_{ij}} \tag{3}$$

ρ_{ij}: Correlation fraction coefficient reflecting the effect of uncertainty source j on X_i
σ_{ij}: Standard deviation of X_{ij}
m_i: Median value of X_i
X_i: Lognormal random variable of basic event i
X_{i0}: Independent impact of X_i
$X_{\cdot j}$: Any one of $X_{1j}, X_{2j}, \ldots, X_{kj}$
i: Basic event
j: Common uncertainty source (j=0: independent effect)

When a random variable, Xi, as shown in Eqn. (1) is assumed to be a lognormal, the probability obtained from the minimal cut sets may be changed by the value of correlation fraction coefficient. Four common uncertainty sources are defined to apply safety culture impact: system, component, failure mode, and department. The vectors for the two basic events among them are:

Basic event 1: (system1, component1, failure mode1, department1)
Basic event 2: (system2, component2, failure mode2, department2)

If the basic event 1 and 2 lies in the same system, both events might have just one common uncertainty source but if their components are also supposed to be same. They will have two common uncertainty sources. The number of common uncertainty source in each minimal cut sets are obtained by analysing the basic events. The ρ_{ij} in the above Eqn. (2) is the degree of common uncertainty source impact on the basic event. If the value of ρ_{ij} is obtained, σ_{ij} is calculated by Eqn. (3). All variables of above Equations are obtained sequentially. It means that the basic events are independent when the score of the safety culture index is 10. For the value of that safety culture index score is 0, it denotes the perfect correlation. On the basis of these assumptions, the Equations to find the value of ρ_{ij} is expressed as follows.

$$\rho_{i0} = \frac{X}{10} \tag{4}$$

$$\rho_{i1} = \rho_{i2} = \rho_{i3} = \rho_{i4} = \frac{10-X}{40} \tag{5}$$

, where the variable, X, is the average of the safety culture index score.

Using the measure of the CDF as shown in the Eqn. (6) below, the Safety Culture Impact Index (SCII) is newly defined.

$$SCII = \frac{CDF(SC)}{CDF} \qquad (6)$$

, where the CDF(SC) means the Core Damage Frequency considering safety culture impact and the CDF denotes the Core Damage Frequency not considering safety culture impact.

3. RESULTS

In order to apply the developed SCII model to the reference nuclear power plant, the minimal cut sets are produced from by running the SAREX code. For the reference plant, the number of the minimal cut sets is a value of 51,212 while the basic events are a value of 1,239. To get a new result of the minimal cut sets considering the safety culture impact, the prototype SCII program using the C# language has been developed in this study. This program might contribute to summarizing and visualizing the safety culture impact for the reference plant. The data shown in Table3 is used and Monte Carlo method is applied to quantify the CDF results using the new minimal cut sets. For the uncertainty analysis, the SCII value provides both the values corresponding to the confidence levels such as 5%, 50%, 95% and the mean value. Figure 1 shows the main screen of the program developed in this study. When the input data is obtained properly and applied in this program, the results are produced in the format shown in Figure 2 which is one of the output displays. The important ones among the outputs include the scores of each safety culture index and the value of SCII. The score of safety culture can be also displayed as the histogram graph and the pie chart. It can be used for comparing each safety culture index of the reference plant. These graphs show the periodic monitoring results and the measures of the SCII changes of the reference plant. The SCII values are also represented according to safety culture indicator score shown in Table 4. The safety culture index score is correlated to the uncertainty of CDF explained above. It shows that the safety culture affects the safety of nuclear power plant quantitatively.

Figure 1: Main screen of the program

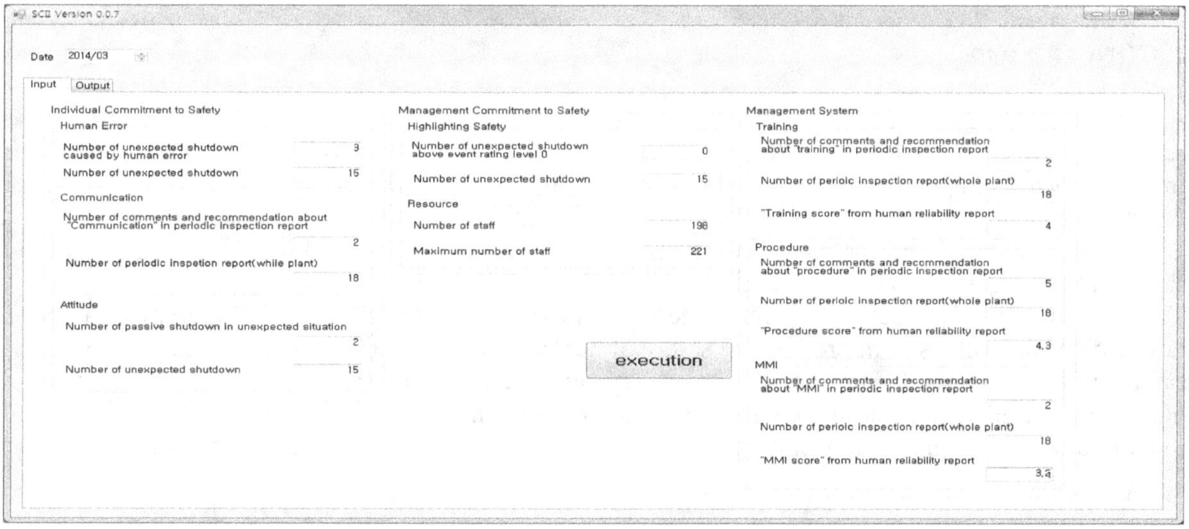

Figure 2: The output screen

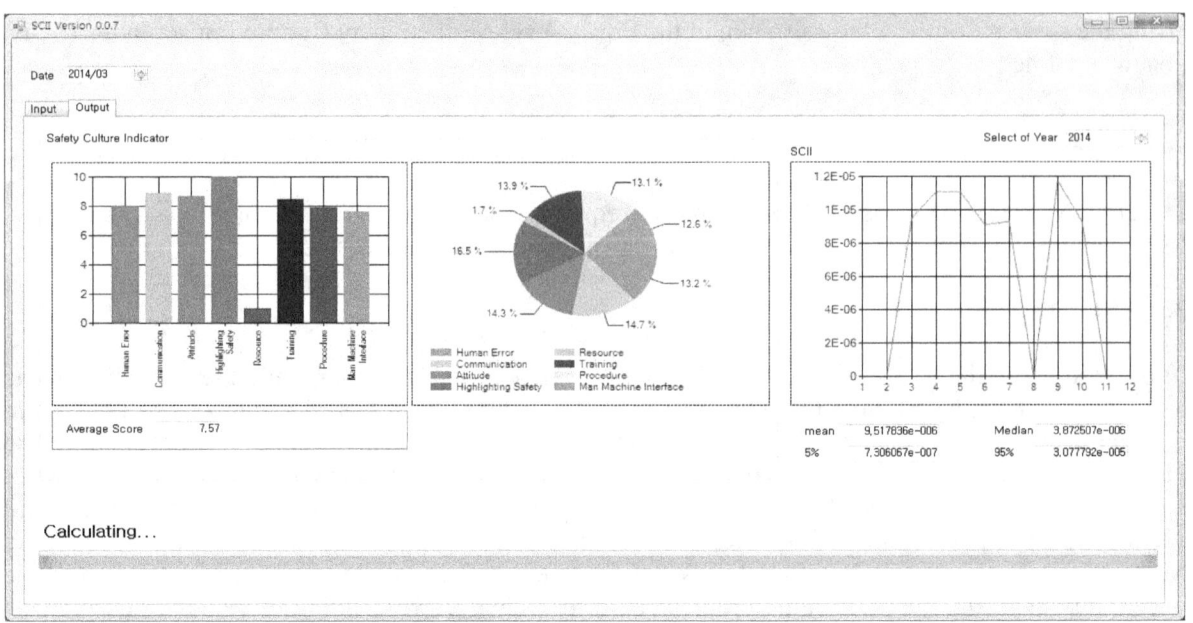

Table 4: SCII of the reference plant

Safety Culture Indicator Score	SCII			
	Mean	5%	50%	95%
0	1.66E+00	7.17E-02	4.81E-01	5.23E+00
2.5	1.43E+00	1.01E-01	5.46E-01	4.48E+00
5	1.22E+00	1.31E-01	6.04E-01	3.89E+00
7.5	1.14E+00	1.95E-01	6.74E-01	3.17E+00
10	1.00E+00	3.21E-01	7.44E-01	2.31E+00

4. CONCLUSION

A new methodology for assessing safety culture impact index has been developed and applied for the reference nuclear power plant. The SCII may contribute to measuring the changes of the core damage frequency which might be affected by the status of safety culture in nuclear power plants. The core damage frequency of accident sequences is obtained by the logical combination of minimum cut sets. The SAREX code is used for producing safety culture impact index related MCS. The uncertainty in safety culture impact has been also analysed.

The developed SCII model might contribute to comparing the level of safety culture among nuclear power plants as well as to improving the safety of nuclear power plants. It is shown that the degree of safety culture affecting the core damage frequency can be estimated. The result of uncertainty analysis may be increased by considering the safety culture impact. The SCII model therefore might contribute to monitoring the level of safety culture and, to improving the safety of nuclear power plants.

Acknowledgements

This work was supported by the Nuclear Power Core Technology Development Program of the Korea Institute of Energy Technology Evaluation and Planning granted financial resource from Ministry of Trade, Industry & Energy, Korea (No. 20131510101690).

References

[1]   International Nuclear Safety Advisory Group. Safety series No.75-INSAG-4 "Safety Culture", International Atomic Energy Agency, Vienna, 1991.

[2]   Keyvan Davoudian, Jya-Syin Wu and George Apostolakis. The work process analysis model (WPAM). Reliability Engineering and System Safety, 45, 107-125, 1994.

[3]   Institute of Nuclear Power Operations. Principles for a Strong Nuclear Safety Culture, 2004.

[4]   Institute of Nuclear Power Operations. Traits of a Healthy Nuclear Safety Culture, 2012.

[5]   U.S. Nuclear Regulatory Commission. Safety Culture Attributes Table, Summary of the Public Meeting on NRC's Safety Culture Initiative, 2005.

[6]   Choi Y.s. et al. Safety Culture Indicators for NPP: international Trends and Development Status in Korea, Proceeding of the Spring Conference on Korea Nuclear Society, 2004.

[7]   KAERI. Development of A Standard Method for Human Reliability Analysis (HRA) of Nuclear Power Plants – Level 1 PSA Full Power Internal HRA -, 2005.

[8]   Qin Zhang. General Method Dealing with Correlations in Uncertainty Propagation in Fault Trees. Reliability Engineering and System Safety, 26, 1989.

[9]   Pusan national university. Development of Indirect Performance Indicator for Performance-Based Operational Framework of Nuclear Power Plants, 2005.

Probabilistic performance assessment for crushing system.
A case study for a mining process.

P. Viveros[a,b], A. Crespo[b], F. Kristjanpoller[a,b], R. Stegmaier[a], E. Johns[a], V. Gonzalez-Prida[b]

[a] Universidad Técnica Federico Santa María, Department of Industrial Engineering, Valparaiso, Chile
City, Country
[b] Department of Industrial Management, University of Seville,

Abstract: The productive performance of a system is mainly determined by its design specifications such as volume, capacity and processing speed; however, it is also conditioned on the reliability of its equipment, the logic be-hind the operation of the process and the availability of its overall system. In this viewpoint, these features are relevant to estimate the throughput, and need to be given due account in proper dimensioning and management.

Significant modelling complexities can arise when accounting for realistic conditions for multi-production, storage flexibility, recirculation, setups, and random times of operations and repairs. Within an integrated, systemic view of the production process and related productivity performance, these issues must be treated by fusing the methods of reliability and availability analyses with those of production process engineering.

This article propose an integrated probabilistic modelling to analyze, evaluate and compare the performance of a Crushing line under specific operational criteria, considering the characteristics of its equipment and the systemic setting in which they are embedded. The resilience characteristic is an important real factor of this kind of process, so will be analyzed in detail.

According to, the software RelPro® will be used to model the Crushing System (mining process in Chile). This software was developed in Java language, based on Monte Carlo simulation (simulation by event). This modelling creates the flexibility needed to model the complex behaviour of high-dimensional systems.

Keywords: System Modelling, Performance Simulation, Simulation by event, Resilience restriction, Primary Crushing Process.

1. INTRODUCTION

In current literature, there are several investigations whose objective is to identify the principal factors that directly affect the maximization of throughput and economic benefit, those that converge at empirical consideration of reliability, maintainability, and availability indicators (RAM). The traditional reliability analyses based on a logical and probabilistic modelling contributes to improve key performance indicators (KPIs) of a system [1], a direct influence in determining optimal operation designs [2]. In this line, there are many alter-natives available for reliability analysis of systems employing analytical techniques, like Markov Models [3], Poisson [4], and other techniques [5]. The systematic study are usually based on techniques like Reliability Block Diagrams (RBDs) [6, 7], Fault Trees (FTs) [8], Reliability Graphics (RGs) [9], Petri Nets (PNs) [10], among others; which allow for the logical relationships that underlie the behaviour or dynamics of the process. In some applications, specifically when complex and dynamic systems are involved, these techniques must be adapted or extended with further considerations. An excellent example for this is the adaptation of de classic RBD to measure the effects of the buffer inventory level on the performances of the production line [11].

In practice, the performance of a production line is limited by intrinsic characteristic of each one of the equipment that contributes to the overall functioning, the most important are:

✓ Nominal Capacity of the machinery/stations/production equipments.
✓ Reliability and Maintainability behaviour

- ✓ Maintenance Planning
- ✓ Operational Restrictions
- ✓ Setting or structure of the system

Their corresponding limitations can create bottlenecks in the production which must be accurately evaluated and effectively corrected [12, 13]. Then, the operational reliability and productivity of a system must be analyzed in a combined fashion to allow optimal exploitation of resources to achieve the set production goals [3]. This requires that a number of characteristics of the production processes be given due account, such as the last mentioned.

In this line, the primary concern of this proposal is to build a model to analyze and project the system performance (mining process) involving realistic criteria last mentioned. This proposal directly derives from industrial requirements in the context of design evaluation.

Monte Carlo simulation is used as the modelling framework to capture the realistic aspects of equipment and system behaviour [15, 16]. This approach creates the flexibility needed to model the complex behaviour of high-dimensional systems.

The most important motivation for using Monte Carlo simulation comes from the possibility of building a realistic (probabilistic) model of a system's (stochastic) behaviour, which allows the creation of realistic system production life representations by sampling the occurrence of discrete random events from their characteristic probability distribution functions. This method is commonly used to solve complex problems by random sampling [17, 18]. It involves the generation of random or pseudo-random numbers that enter into an inverse probability distribution, resulting in as many scenarios as the number of simulations made [19]. The results of this process being far more informative than what can be inferred from a few designed scenarios, e.g. generated for 'what if' type analyses.

In this paper a Monte Carlo simulation-based analysis procedure is used to analyze a real-world case study from mining engineering. The simulation model will be implemented in the RelPro environment [20], estimating the expected behaviour of performance of each piece of equipment and of the system as a whole, and generates related confidence bounds that account for the statistical variability in behaviour.

RelPro is an analysis and simulation tool that can be used to model continuous and discrete production systems, such as conveyors, transfer lines, mass production lines, fleets, and others. RelPro allows the reproduction of randomized replications of a system model using highly complex logic and it provide innovative and efficient algorithms to analyze and evaluate different scenarios, supporting making decision process related to design and operational conditions, aiding of course the business result.

The motivation of this work is to build an integral probabilistic modelling for a mining process (Crushing line), which constitutes a systematic procedure to model, simulate and sensitize the selected production process, all under innovative algorithms and friendly RelPro environment.

According to the aims, this article is organized as follows: in section ''System Description,'' the application is presented in detail; in sections ''Modelling of the system'' the process is modeled under RelPro environment and briefly summarized according to the general methodology; in section ''Data Analysis,'' will be explained the importance of the data and reliability and maintainability analysis with RelPro.

Finally, case study is solved in section "Simulation Model" and some concluding remarks are given in section ''Conclusion''. .

2. SYSTEM DESCRIPTION

In the context of mining industry, this paper presents and analyses a real case study developed in a cooper Open pit mine, specifically for the primary crushing (PC) (Fig. 1), which normally is the first stage in a comminution process [1]. Crushing is normally carried out on 'run-of mine' ore, and the objective is to reduce the size of the material from the mine, which is then transported by some conveyor belts to a stockpile.

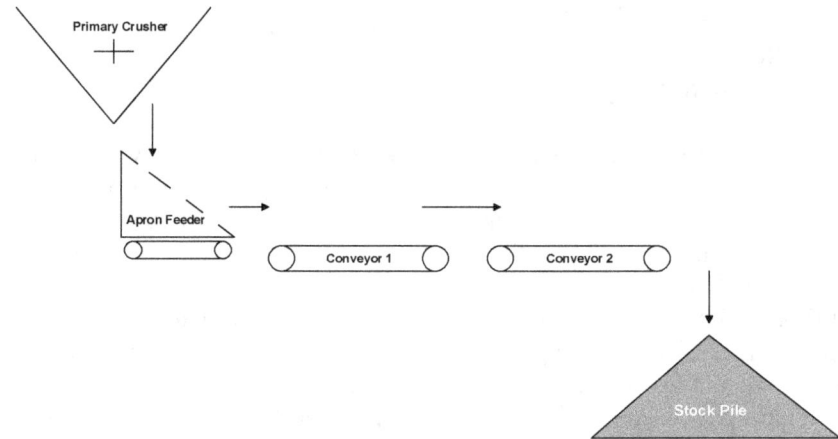

Fig. 1 Process diagram for the primary crushing process

As a brief description of the process involved, after a mining company has removed overburden, extraction of the mineral ore begins using specialized heavy equipment and machinery, such as loaders, haulers, and dump trucks, which transport the ore to processing facilities using haul roads. After, the ore is dumped into the primary crusher; then an apron feeder is connected controlling the gravity flow of bulk solids, providing an uniform feedrate to the next receiving belt conveyor. Two next belt conveyors are connected to the apron feeder, to finally feed the stockpile.

The main characteristics of the primary crushing process shown in Fig. 1 are listed in Table 2.

Table 1. Primary crushing process information

Equipment	ID	Basic Fucntion
Primary Crusher	CH_001	Mineral size reduction
Apron Feeder	FEED_001	Control of the gravity flow of bulk solids, providing an uniform feedrate to the next receiving belt conveyor
Conveyor Belt 1	CONV_001	Transport the crushed mineral to the next conveyor
Conveyor Belt 2	CONV_002	Transport the crushed mineral to the stock pile

2. MODELLING OF THE SYSTEM

The logic behind the operation (functional dependency) of the process can be understood by using a simple question What' if? It means that it is necessary to recognize the effect of some random or planned state change of any production equipment/machinery of the process over the system, that involve the effect in terms of functioning and work load capacity over the others machineries, subsystems and overall system. Normally, there are two possible states, degradation (normal established functioning) and not degradation (failure state, preventive intervention or operational detention) [21].

The four components of the process are connected in a simple serial setting, which implies that any single failure will cause the entire system to fail. A major operational criteria that benefits the outcome (second scenario to model and sensitive) is the resilience of the process when the primary crusher or the apron feeder fails. When one fails, or both simultaneously, the downstream process will continue to work for the next 40 minutes. This operational feature is equivalent to if both machineries have the ability to accumulate material during normal operation, been capable to supply 30 to 40 minutes of downstream operation

The resilience scenario leads to a cold standby system [22], which satisfies the usual conditions (i.i.d. random variables, perfect repair, instantaneous and perfect switch, queueing). It is important to consider tree important features:

- ✓ To model it, it is necessary to create a "virtual" stand by equipment, with specific parameters of failure and repair.
- ✓ As preliminary criteria, the failure distribution must be a Uniform Distribution with parameter of life equal to range of the resilience time estimated (30 – 40 minutes).
- ✓ As preliminary criteria, the repair time distribution of the "virtual" equipment must be equivalent to the repair time distribution of the main equipment. It is a conservative scenario.

The Fault tree diagrams are developed (Fig. 2 and 3) to support the understanding and representation of the both process scenarios.

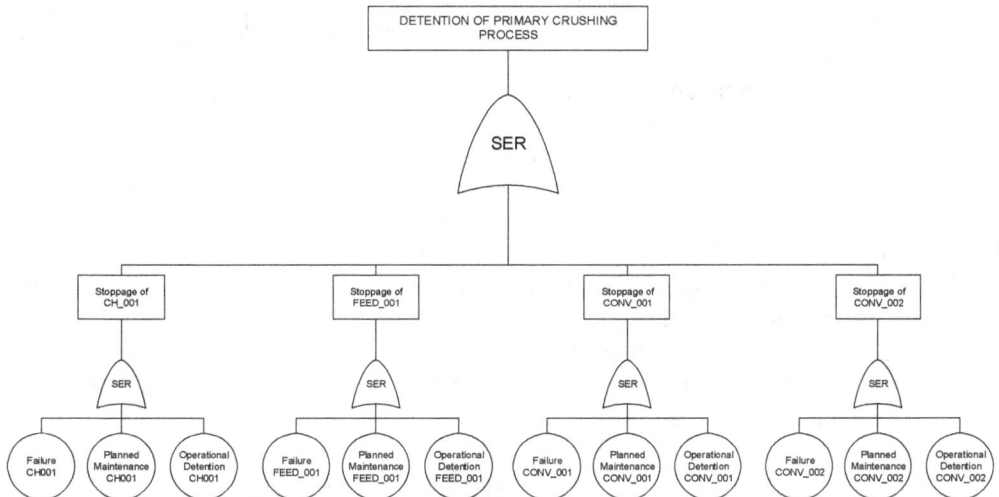

Fig. 2 FT representation of the primary crushing process - immediate effect

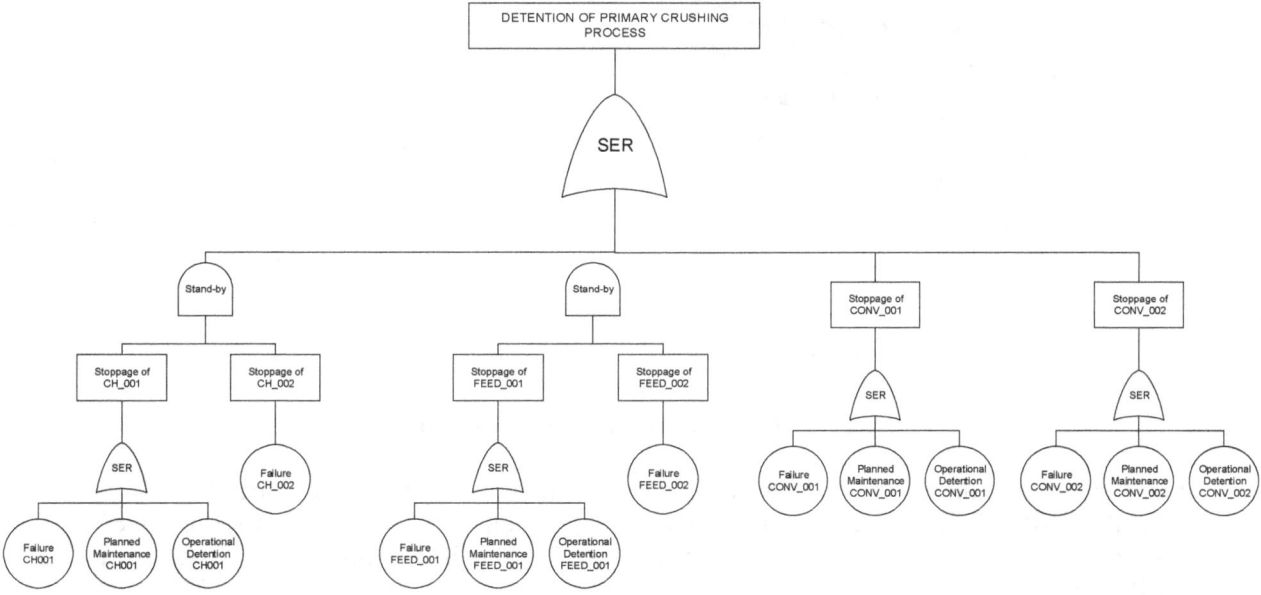

Fig. 3 FT representation of the primary crushing process – resilience approximation

So, process modelling in software RelPro must consider the traditional scenario (immediate effect of detention) and the constraint scenario (resilience approximation). With this, the analysis results will be enriched.

As was indicated at the beginning of the paper, the motivation of this work is to build an integral probabilistic modelling, so the next section will explain and analyze the statistical data related to: Times To Failure (TTF) associated to reliability and Time To Repair (TTR) related to maintainability.

The simulation will not include parameters linked to operational stoppages nor planned maintenance. This consideration just simplified the analysis in terms quantity of analysis, but not in terms of quality or methodology, since these considerations can be modelled and integrated just like a serial setting as was graphically represented by the FT diagrams (Fig. 2 and 3).

3. DATA PARAMETERIZATION

The definition of the probability distributions is commonly used to describe the failure and repair processes of the equipment. Different types of statistical distributions are examined and their parameters are estimated by using, as mentioned before, the RelPro Application. The software fits several distribution models based on the historical data, and it is possible to choose and use a preferred model, or accept the distribution recommended by the software (Weibull 2 parameters, Exponential, Lognormal, Normal, Dirac Delta and Uniform).

The following step in data management is to determine the nature of the equipment involved in the process, so the distributions must be selected under relevant stochastic models, according to the behaviour of the data in terms of trend and independence.

Analyzing the historical data of the equipment involved, independence and trend indicators are calculated. In the first instance, this feature is observed graphically. For this, some graphics of cumulative time to failure (TTF) observe the behaviour of trends and then dispersion charts of successive lives to observe the degree of correlation of variables or independence. Also, the Laplace test was applied. Due to space limitations, these are not included.

The Software RelPro allowed to estimate all the parameters for each probability density function (TTF and TTR), and no trend was identified. As an example, Fig. 4 shows the parameterization for the primary crusher, specifically for times to failures (TTF).

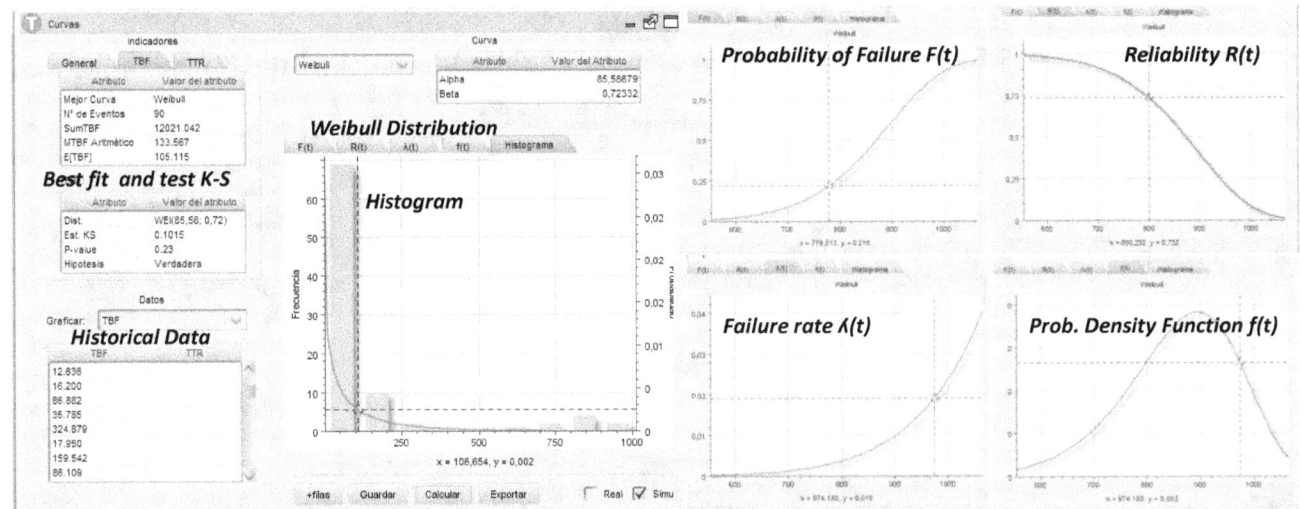

Fig. 4 Probability density function for primary crusher

Fig. 4 summarize the information about: histogram of failure, Accumulated probability of failure F(t), reliability R(t), failure rate Λ(t), probability density function of failures f(t) and the relevant information about the Kolmogorov–Smirnov tests [23] (statistical goodness-of-fit test selected in RelPro).

Table 2 summarizes main parameters and key indicator related to reliability and maintainability.

Table 2. Reliability and maintainability information

Equipment	Time To failure Parameterization				Time To Repair Parameterization			
	Best fit Distribution	Parameter 1	Parameter 2	MTBFi	Best fit Distribution	Parameter 1	Parameter 2	MTTRi
CH_001	Weibull	α=85,72	β=0,72	106	Normal	μ=4,1	σ=1,12	4,10
FEED_001	Weibull	α=82,01	β=0,87	88	Normal	μ=3,9	σ=1,31	3,90
CONV_001	Exponential	λ=0,054		19	Normal	μ=1,2	σ=0,60	1,20
CONV_002	Weibull	α=15,84	β=0,65	22	Exponential	λ=0,76		1,32

4. SIMULATION MODEL

To model and simulate the process, it is necessary to consider all the specific operating conditions and all realistic restrictions that exist and are physically respected by the real process. The main characteristics of each piece of equipment to be considered in the simulation model are listed in Table 2, and the restriction related to the logical and functional dependency were explained in detail in the section: Modelling of the system. The FT for both main scenarios helps to build the model in RelPro environment.

The simulation must consider an overall production rate, which is similar for all equipment according to the serial setting explained. Each piece of equipment must be able to produce at the rate required by the process, this being totally or partially as demanded by the system.

For this specific case, the rate considered is 3000 ton per hour, and it assumes that the ore input is equivalent to the ore rate output demanded by the process. This means that in any case the system will stop for lack of supply or for capacity problem after the second conveyor belt (feeding the stockpile).

The graphical models (base for the simulation) developed in RelPro are presented and discussed next.

4.1. About RelPro

Processing systems depends in part on the operating logic established, for this RelPro has efficient algorithms dedicated exclusively to the representation and analysis of these logics. Most of continuous simulators, or discrete but with continuous control and monitoring variables, perform the calculation of indicators and identification of states through monitoring at certain intervals of time (usually very small), this procedure is slightly efficient when it is compared to vision oriented just to the state change of components of the system. That means that the monitoring and consultation is performed only when something in the system changes state, either random or planned condition. For this, a continuing evaluation of the state of each system element is required. So, in the field of simulation, RelPro is a simulator based on discrete-event occurrence, in contrast with continuous simulation in which the simulation continuously tracks the system dynamics over time. The impact generated depends exclusively on the established functional dependencies and diagrammed in RelPro environment.

The main elements of the modeling are: Tree of components representing the hierarchical structure in the systems, and the flow chart includes:
  ✓ Actionable nodes representing systems, subsystems or equipment, logic-nodes configuration (method by which distributed or flow conditions over the subsequent process) input and output nodes (clarifies the input and output of material processed).
  ✓ Bows, correspond graphically to arrows, represent the transfer of flux.

The graphical models (base for the simulation) developed in RelPro are presented and discussed next

4.2. Simulation Modelling and Analysis in RelPro environment

Now, as was mentioned at the end of the section "Modelling of the System" RelPro will consider the traditional scenario (immediate effect of detention) and the constraint scenario (resilience approximation). With this, the simulation modelling is:

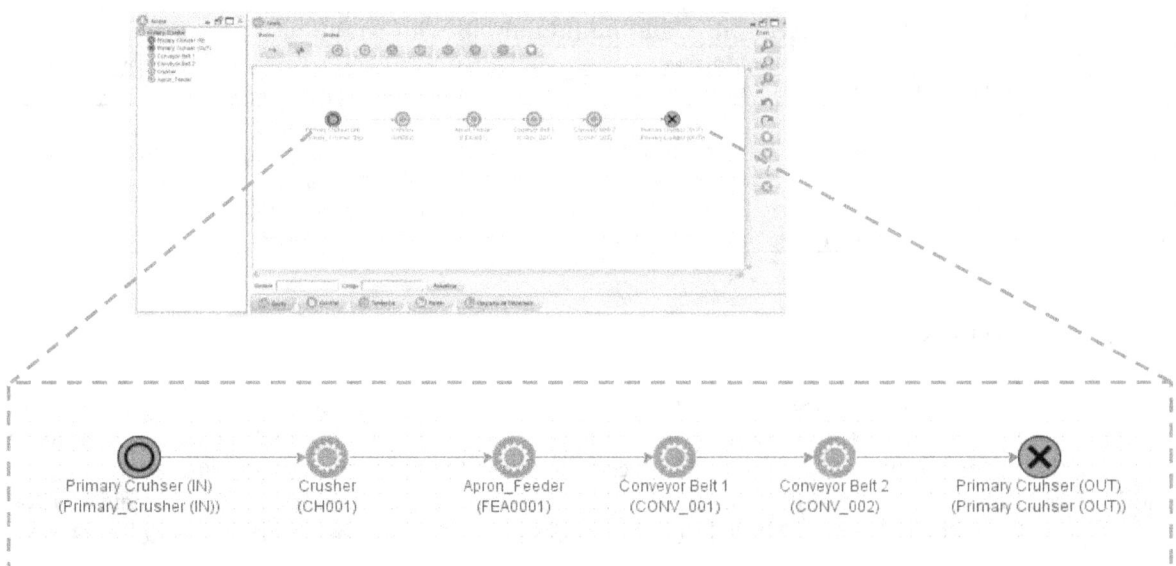

Fig. 5 Graphical representation of modelling in RelPro environment – Immediate effect scenario

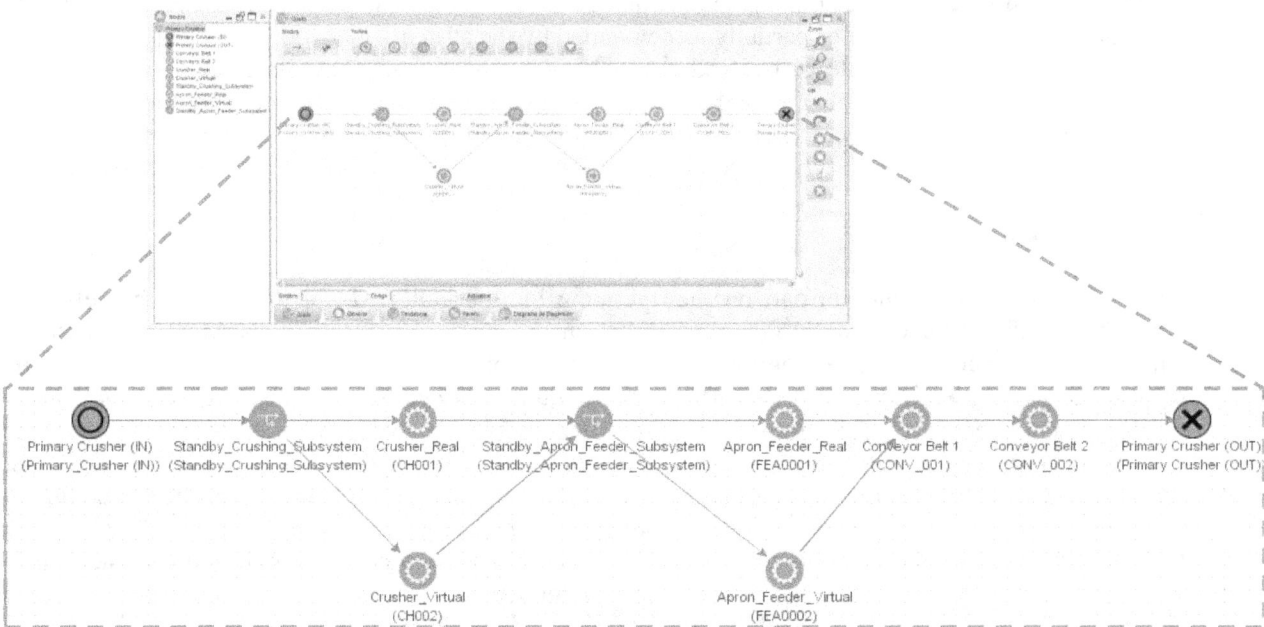

Fig. 6 Graphical representation of modelling in RelPro environment – Resilience approximation scenario

For both scenarios is required (inputs) the data about the characteristics of each piece of equipment considered in the simulation (See table 2). Furthermore, for "resilience approximation" it is assumed that the standby equipments (virtual machineries for modelling) come into operation immediately after the failure of the primary machinery (Crusher and Apron Feeder) and the repair actions are independent. This consideration is traditionally recognized as cold-standby [22].

As we know, the resiliencie time for primary machineries is between 30 and 40 minutes, so the parameters of life degradation and repair time will be modelled by Uniform Distribution..

4.2. Simulation Results

A total of 1000 replication were performed over a time horizon of 1 year (8760 hours) of operation. The main reason for selecting this simulation horizon, executed 1000 times, is to provide a representative sample to generate histograms readable and compelling indicators. In addition, some pieces of equipments have small times to failure values (e.g. Belt Conveyors); so on the time horizon will become very significant. The results of the 1000 simulation are summarized in table 3 and table 4. The performance indicators to measure are: Mean % Availability, Mean % of Operation, MTTF, MTTR and the Mean of total production of the sys-tem.

The results for the immediate effect scenario are:

Table 3. Summary of simulation results – Immediate effect scenario

Equipment	Indicator of Performance				
	Mean % Availability	Mean % Oper. Time	Mean Production (MM Tons)	MTTF	MTTR
CRUSHING SYSTEM	81,25%	81,25%	21,355	8,24	1,90
CH_001	96,10%	81,25%	21,355	90,41	4,11
FEED_001	95,72%	81,25%	21,355	75,06	3,93
CONV_001	93,75%	81,25%	21,355	16,10	1,23
CONV_002	94,18%	81,25%	21,355	18,56	1,32

According to the results and in relative terms, CONV_001 and CONV_002 will be the critical equipment in terms of availability (93,75% and 94,18%). The expected mean production of the system is 21,355 Million of Tons, equivalent to 7.118,17 hours of operation. As the model simulation does not include planned stoppages (maintenance or operational stoppages), the % mean availability of the crushing system is equal to the % mean operational (81,25%).

Another important result, from the systemic point of view, is the frequency of failure which is each 8,24 hours of functioning, and the mean time to repair is around 1,9 hours. These last indicators are the main reason of the low % mean operational time, mainly represented by the high frequency of failure of the system. As the logical configuration is in series, any change state (planned or not planned) of any equipment will impact over the change state of the overall system.

So, to improve the reliability of the overall process (decrease the frequency of system failure) will be necessary to improve the reliability of conveyors CONV_001 and CONV_002, this means increasing the mean times to failure, 16,10 and 18,56 respectively.

A direct analysis of maintainability indicators suggest that we should not be concern about it, however, if the direct cause of the reliability results of single equipments is the low quality of maintenance execution (e.g. poor technical skills of maintenance personnel, spare parts in poor condition, lack of work procedures, environmental conditions, and other.), efforts should be focused to correct deviations in reliability and maintainability.

Next will be presented the histogram of Production (Tons). The histograms for availability and % of Operational time can be obtained directly from the Software RelPro.

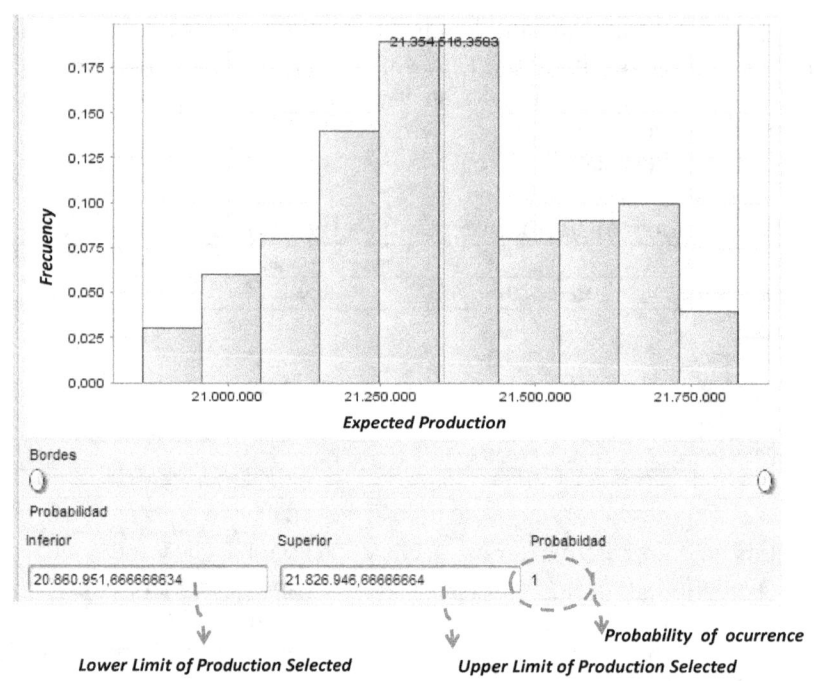

Fig. 7 KPI´s Histograms for Immediate effect scenario

Table 4. Summary of simulation results – Resilience approximation scenario

Equipment	Indicator of Performance				
	Mean % Availability	Mean % Oper. Time	Mean Production (MM Tons)	MTTF	MTTR
CRUSHING SYSTEM	82,53%	82,53%	21,689	8,42	1,78
STANDBY PRIMARY CRUSHER SUBSYSTEM	96,89%	82,53%	21,689	93,44	3,53
CH_001	96,29%	82,53%	21,568	92,85	4,12
CH_002	14,56%	82,53%	0,012	0,59	4,11
STANDBY APRON FEEDER SUBSYSTEM	96,43%	82,53%	21,689	76,63	3,29
FEED_001	95,79%	82,53%	21,547	76,04	3,88
FEED_002	15,19%	82,53%	0,142	0,59	3,91
CONV_001	93,76%	82,53%	21,689	16,36	1,23
CONV_002	94,25%	82,53%	21,689	18,87	1,30

Again, conveyors are the critical equipment in terms of availability. The primary crusher and Apron feeder subsystems have increased their availability thanks to the virtual equipments configured into the RelPro environment. The mean production is 21,689 Million of Tons, equivalent to 7.229,66 hours of operation. Similarity to the previous scenario simulated, the model simulation does not include planned stoppages (maintenance or operational stoppages), so the % mean availability of the crushing system is equal to the % mean operational time (82,53%). The latter is a key indicator to compare the results between simulation models. The results of frequency of failure (8,42 hours of functioning) and mean time to repair (1,78 hours) also have improved, supporting the increased production (+ 0.3 million of tones) and availability (+1,3%) results.

Next will be presented the histogram of Production (Tons). The histograms for availability and % of Operational time can be obtained directly from the RelPro environment.

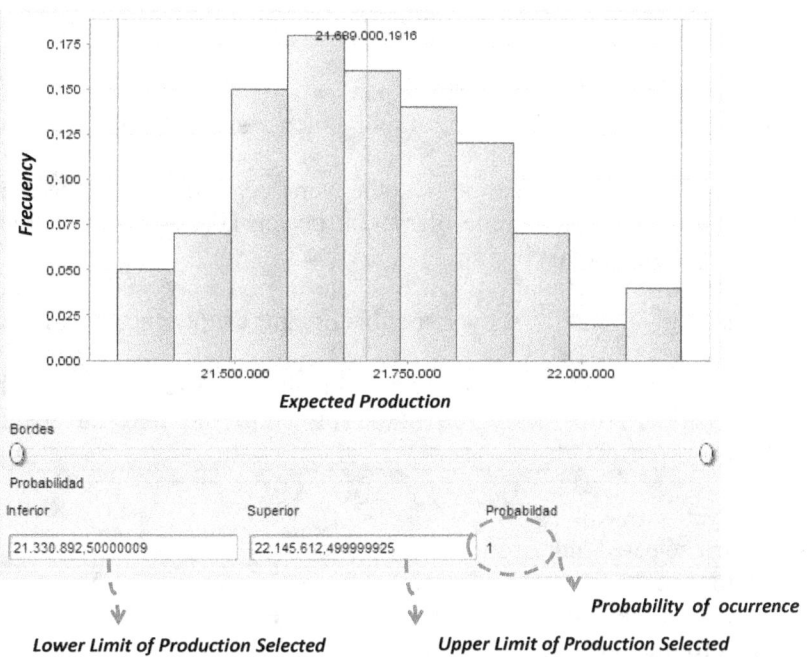

Fig. 8 KPI´s Histograms for Resilience approximation scenario

The variability in the production level shown in Fig. 7 and 8 is due to the stochastic characteristic of the behaviour of the equipments in the system. Also, the convergence and concentration of area around the aver-age of the production histogram supports the good results obtained with 1000 simulations.

As a special case, the % mean availability for virtual equipment (CH_002 and FEED_002) is calculated considering that the time horizon for the calculation is only during the primary equipment repair, so this percentage represent the mean % of time where the virtual equipment support to the primary equipment, and its equivalent to 14,5% and 15, 19% respectively.

Comparing the results of the simulated scenarios, it can be concluded:
- ✓ The considered resilience contributes significantly to the outcome of the business, validated by the increased availability, operation time and the expected production.
- ✓ The Standby approximation modeled in RelPro meets the objectives pursued by analysts.
- ✓ For both, the main problem is the reliability of the selected critical equipment, this because of the high frequency of failure. So, next research must be focused to identify the primary causes of high frequencies trough, e.g. root cause analysis [24].
- ✓ Maintainability is controlled and requires no further attention since the focus of improvement is the reliability.

5. CONCLUSION

Performance analysis must be an integral part of mine engineering assessment and operational management, controlling operating plants or evaluating new designs project. Simulation is powerful tool to estimate performance (design stage), even more when characteristics of reliability, maintainability, productivity and functional dependencies features are integrated to the model. The main result of this paper is a new modelling approach to simulate a production plant, developing a case study of a real mining process (primary crushing process), including a specific scenario with a restriction formally known as resilience. It was implemented via the simulation program RelPro.

The numerical results clarify the effect of the resilience in the performance results (1,3% increase in availability and production) and allows preliminarily identify critical equipment or possible bottlenecks, in terms of reliability and maintainability. The detail of results are clearly specified and explained in last section.

As a summary, the result of the modelling allows:
- ✓ Project the performance of each piece of equipment, subsystems, and overall crushing systems.
- ✓ It is possible to identify the equipment (s) with the worst performance – Potential bottlenecks.
- ✓ Identify responsibilities in the outcome of system performance, acknowledging directly the effect of reliability and maintainability.
- ✓ With the histograms of the simulation will possible to make a decision with a level of risk (probability), e.g. Fig. 7 and 8 shows the histogram of production and the respective probability.
- ✓ Compare the result for both scenarios, calculating the expected effect of the operational restriction (resilience). Furthermore, for future research, the time of resilience may be sensitized and evaluated if necessary.

Future possibilities to analyze with RelPro:
- ✓ Histograms for each selected indicator of performance.
- ✓ Add new indicator, such as: number of failure events, preventive events and operational detention events; total time of corrective maintenance/preventive maintenance/operational detentions; budget for maintenance, and others.
- ✓ Basic cause of the Operational stoppages, it refers to intrinsic detention of the equipment (e.g. misalignment of the conveyor belts) or Operational stoppages propagated from other piece of equipment in the system (e.g. if the belt conveyor 1 fail the rest of the system will stop obligatory. So, this event will be recorded as a detention propagated in the rest of equipments of the system).
- ✓ The modelling method may be adopted in order to analyze more complex systems or process.

Future possibilities to sensitize and analyze with RelPro:
- ✓ Probabilistic parameters of life and repair (genetic).
- ✓ Preventive frequencies at equipment level.
- ✓ Design of the process, involving redundancies, priorities, load sharing and overload capacity. Furthermore, recirculation characteristics.
- ✓ Time of resilience (increase or decrease) and evaluate the impact evolution.

Finally, the authors encourage the use of this model to evaluate the expected performance as early as at the design stage, ensuring highly efficient investments and positive impacts on future productivity.

REFERENCES

[1]   Viveros, P., Zio, E., Kristjanpoller, F. and Arata, A. Integrated system reliability and productive capacity analysis of a production line. A case study for a Chilean mining process. Proc. IMechE, Part O: J. Risk Reliab., 2012; vol. 226, 3: pp. 305-317.

[2]   Gharbi, A., Kenne´, J.-P., and Beit, M. Optimal safety stocks and preventive maintenance periods in unreliable manufacturing systems. Int. J. Prod. Econ., 2007, 107, 422–434.

[3]   Buzacott, J. A. and Shanthikumar, J. G. Stochastic models of manufacturing systems, 1993 (Prentice- Hall, Englewood Cliffs, New Jersey).

[4]   Meller, R. D. and Kim, D. S. The impact of preventive maintenance on system cost and buffer size. Eur. J. Oper. Res., 1996, 95, 577–591.

[5]   Huang, D. and Billinton, R. Impacts of repair state residence time distributions in an electric power generating capacity adequacy assessment. Proc. IMechE, Part O: J. Risk Reliab., 2007, 221, 297–305.

[6]   Zio, E. and Pedroni, N. Building confidence in the reliability assessment of thermal-hydraulic passive systems. Reliab. Engng Syst. Saf., 2009, 94(1), 268–281. 25 Zio, E., Podofillini

[7] Zio, E., Podofillini, L., and Zille, V. A combination of Monte Carlo simulation and cellular automata for computing the availability of complex network systems. Reliab. Engng Syst. Saf., 2006, 91, 181–190.

[8] Marseguerra, M. and Zio, E. Basics of the Monte Carlo method with application to system reliability, 2002 (LiLoLe-Verlag GmbH, Hagen, Germany).

[9] Crespo, A., Sánchez, A., and Benoit, L. Monte Carlo based assessment of system availability. A case study for cogeneration plants. Reliab. Engng Syst. Saf., 2005, 88, 273–289.

[10] Zio, E. and Pedroni, N. Reliability estimation by advanced Monte Carlo simulation. In Simulation methods for reliability and availability of complex systems. Springer Series in Reliability Engineering, 2010, Part I, 3-39, DOI: 10.1007/978-1-84882-213-9_1.

[11] Macchi, M., Kristjanpoller, F., Garetti, M., Arata, A., Fumagalli, L. Introducing buffer inventories in the RBD analysis of process production systems. Reliab. Engng Syst. Saf., 2012, 104, 84–95.

[12] Goldratt, E. The goal: a process of ongoing improvement, 1992 (North River Press, Great Barrington, Massachusetts).

[13] 16 Roser, C., Nakano, M., and Tanaka, M. A practical bottleneck detection method. The Winter Simulation Conference, 2001.

[14] Zio, E. and Pedroni, N. Building confidence in the reliability assessment of thermal-hydraulic passive systems. Reliab. Engng Syst. Saf., 2009, 94(1), 268–281.

[15] Zio, E., Podofillini, L., and Zille, V. A combination of Monte Carlo simulation and cellular automata for computing the availability of complex network systems. Reliab. Engng Syst. Saf., 2006, 91, 181–190.

[16] Marseguerra, M. and Zio, E. Basics of the Monte Carlo method with application to system reliability, 2002 (LiLoLe-Verlag GmbH, Hagen, Germany).

[17] Metropolis, M., Ulam, S., 1949. The monte carlo method. Journal of the American Statistical Association 44 (247), 335–341.

[18] Sobol, I.M., 1994. A primer for the monte carlo method. CRC Press, Boca Raton, FL, pp. 107.

[19] Juan P. Vargas, Jair C. Koppeb and Sebastián Pérez. Monte Carlo simulation as a tool for tunneling planning. Tunnelling and Underground Space Technology 40 (2014) 203 – 209.

[20] RelPro. RelPro. Reliability Engineering & Asset Management. www.relproengineering.com; March 2014.

[21] Book Section D. Engineering Asset Lifecycle Management. A review on degradation models in reliability analysis. Gorjian, Nima; Ma, Lin; Mittinty, Murthy ; Yarlagadda, Prasad and Sun, Yong. P 369-384, 2010. Springer London.

[22] A. Birolini. Quality and Reliability of Technical SystemsSpringer, Berlin (1994).

[23] Stephens MA. EDF statistics for goodness of fit and some comparisons. Journal of the American Statistical Association 1974; 69:730–7.

[24] Viveros, P., Zio, E., Nikulin, C., Stegmaier, R. and Bravo G. Resolving equipment failure causes by root cause analysis and theory of inventive problem solving. Proc. IMechE, Part O: J. Risk Reliab., 2014; vol. 228, 1: pp. 93-111.

An innovative proposal for systemic modeling, analysis and simulation in a continuous production process

René Tapia[a*], Pablo Viveros[b,c], Adolfo Crespo[c]

[a]RelPro S.A, Santiago, Chile
[b]Universidad Técnica Federico Santa María, Department of Industrial Engineering, Valparaíso, Chile
[c]Department of Industrial Management, University of Seville, Spain

Abstract:

This research aims to develop an innovative proposal for systemic modeling, analysis and simulation. There are different techniques to estimate availability and production in continuous production plants, being one of the most used the Reliability Block Diagram technique [1-3] because of its simplicity and probability framework, but it has significant approximations. Another widely used technique to model and analyze are Markov-Chains (discrete time) [4]. Both techniques have been specially designed to analyze for a limited number of functional settings and disposition of elements, like parallel, stand-by, serial, and so on [5-7]. Additionally, the last mentioned techniques do not allow studying the variability of variables which is commonly demanded.

So this paper is focused to develop a technique based on specific algorithms which allows engineers to have a broad perspective of the system with a flexible layout framework, simulating the production, availability and runtime of the plant considering the impact of each random event over each one of all elements of the process. The presented algorithms have the advantage that are made based on the occurrence of each event, so the time duration of processing is based on the number of random events [8, 9], reducing dramatically the simulation times.

Keywords: Availability Analysis, Reliability, Simulation of Maintenance, Continuous Production Plants, RelPro

1. INTRODUCTION

The study of the behavior and performance of production systems have been always a constant need, in order to get a maximum return on the basis of costs and expenses to be taken. To estimate the production based on maintenance activities, i.e. repair or other kind of events, formerly was not a task to be executed in design stages, but nowadays is an essential process towards competitiveness.

Diverse methodologies have been developed to measure the performance of the plants or systems, like Markov Chains, Petri-Nets and the extension of Reliability Block Diagrams applied to estimate availability. And more recently has been incorporated the computer simulation of the process. Each methodology has its own advantages and disadvantages, which will be studied further.

The main disadvantage of the mentioned methodologies is that most of them do not consider the dynamic impact of the events that occurs in any equipment to each one of the rest of the equipment of the plant. Markov Chains and Petri-Nets are technically impossible to implement at a level of the entire plant, because of the large number of equipment, and even in computers the layout of the states becomes technically infeasible. Reliability Block Diagrams constitute just an approach, which can lead to significant errors. And most of the simulation environments are not designed to model and represent the impact of the events of each equipment to each one located in the plant.

2. PROBLEM STATEMENT

Current methodologies presents several disadvantages in modelling and representation of production systems, the most important disadvantages are:

*rene.tapia@relpro.pro

- Most the methodologies are approaches.
- Most of the methodologies do not consider risk
- There is no methodology that considers the Life Cicle of the assets in the estimation.

2.1. Approaches to estimate

There are some important approaches in the estimation and calculus of availability in productive systems. One of the main approaches, with higher impact from the theoretical vision to a real vision is related mainly with the use of Markov Chains, assuming that the transition rates are constant, and during all the calculated period. Many authors in this specific field of availability and production calculation uses Markov Chains [4, 5, 6], but indeed they only use constant transition rates, which means only exponential distributed times [10]. However, processes of interest like reliability, availability, maintainability, and safety (RAMS) do not necessarily suggest a behavior described by exponential times. In this matter Petri-Nets have is advantaged because this methodology, in spite of being a formalization of transitions and occurrence of events in a single graphical modelling, is used in simulation mainly in Generalized Stochastic Petri-Nets (GSPNs) [11], in which allows to establish customized time probability distributions, (this will be studied further in next sections). In the case of RBD, there is also possible to use custom probability distributions of time [1]. But PNs (GSPNs incl.), Markov-Chains and RBD are also not allowed to do analysis in production systems in which there are machines that work by default with idle capacities, unless significant changes are made in the modeling, that in the case of RBD and Markov-Chains are necessarily approximations. Other approach is that those methodologies does not consider the different production levels that the system could have, are simplified to be used with a single production rate.

In the special case of RBD, it has several approaches, especially in the calculation of availability of serial systems, not considering the propagation or impact of events of a specific machine in the rest of the machines, this is clear when is known that the availability of serial systems is the product from the availability of each of the machines, calculated previously independently of the behavior of the others.

2.2. Risk consideration

The evaluation of the average performance of manufacturing systems has been widely investigated in the manufacturing system engineering literature. However, there is industrial evidence that production variability due to random disturbances cause the observed production rate to be different from its average value [14]. Analyzing the variability is possible to find optimal designs, maintenance policies, and so on that meet desired service levels. None of the most used methodologies has been deeply developed, or have been expanded with the focus on the study of the risk and variability of the indicators of interest, production and availability. The mentioned methodologies, PNs, Markov-Chains and RBD are completely focused on the study of the mean and expected values of the key performance indicators. There are not total complete research in risk using those methodologies, there are just simplified mechanisms and as results of numerical examples, we can highlight the work of Xin-Feng et al [12], were discrete production is studied, and Manitz et. al [13] and Li S. et al[18] in the continuous case. One the few works that studies the variability is Li C. et al [15], but using discrete simulation. Colledani et al. [14] also presents a methodology of decomposition to obtain the total variability of a line with buffers, but is only applicable for series systems.

2.3. Life Cicle consideration

Until recent time the calculus of reliability was only considering the models described by Perfect Renewal Process, it means, modeling that consider the situation "as good as new" once the maintenance activity over the asset has been done. During the last years other modelling considerations have been adopted with more presence in reliability research, like Non-Homogeneus Poison Process (NHHP) [20] and the Generalized Renewal Process (GRP) [16], in example [20, 21, 22, 23]. But these research works are not focused in evaluate the trend of availability and production of an equipment, this means that are far away to evaluate the trend of availability and production of a system.

None of these methodologies currently allow to implement a long-term analysis unless some special module is developed, or discrete calculations are carried out under different system conditions. While considering the above citations is possible to calculate the expected availability over time to an equipment, it is difficult to do for a complete system. Finally it is concluded that there is no methodology that appropriately consider the equipment degradation over time in a systemic environment to estimate the availability and production.

3. STATE OF THE ART

Currently there are some techniques and methodologies to estimate availability and production in continuous production systems. The most used ones are Markov Chains, PetriNets, Reliability Block Diagram and Simulation Softwares. We proceed to describe these methodologies below:

3.1. Markov Chains

There are discrete time and continuous Markov chains, for purposes of analysis of continuous production systems we consider homogeneous continuous time Markov chain (CTMC) [11]. CTMC is a state space model, in which each state represents various conditions of the system. In homogeneous CTMCs, transitions from one state to another occur after a time that is exponentially distributed. The arcs representing a transition from one state to another are labeled by the constant rate corresponding to the exponentially distributed time of the transition. If a state in the CTMC has no transitions leaving it, then that state is called an absorbing state, and a CTMC with one or more such states is said to be an absorbing CTMC. For the multiprocessor example, we now illustrate how a Markov chain can be developed in a simple way for 2 machines.

Below is developed an example of Markov chains analysis implementation, a productive subsystem comprising two identical devices in parallel is developed.

Figure 1. RBD diagram for Markov chain analysis example.

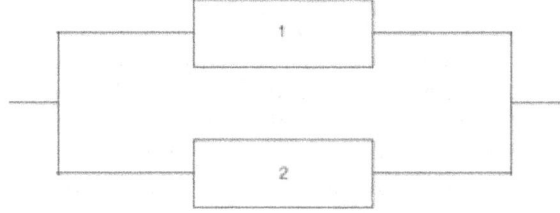

There are 3 cases only, given that the equipment are identical. It is defined $P_i(t)$ as the probability that the systems is in the state i at the time t. In addition has the boundary condition that at the beginning both units are operating normally, this implies that:

$$P_0(0) = 1$$
$$P_1(0) = 0 \quad\quad\quad (1)$$
$$P_2(0) = 0$$

Figure 1. Markov chain diagram for 2 devices in parallel.

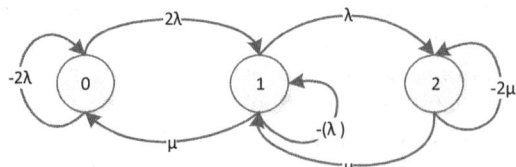

It is possible to omit the differential analysis and move immediately to the steady state analysis, when probabilities $P_i(t)$ are invariant in time. Now the equations that remain are, including the boundary conditions:

$$-2\lambda P_0 + \mu P_1 = 0$$
$$2\lambda P_0 - (\lambda + \mu)P_1 + \mu P_2 = 0$$
$$\lambda P_1 - \mu P_2 = 0$$
$$P_0(0) = 1; \; P_1(0) = 0; \; P_2(0) = 0$$

(3)

Based on the presented system solving for $P_0 + P_1$ corresponds to the availability of the system. In this case is:

$$A = P_0 + P_1 = \frac{\mu^2 + 2\lambda\mu}{2\lambda^2 + \mu^2 + 2\lambda\mu} \blacksquare$$

(4)

Based on the example the difficulty level that may exist when it comes to the analysis of a more complete system is seen clearly. Therefore, it becomes practically an infeasible methodology to more complex systems. Also, consider constant rates of occurrence of events transforms Markov chains in an unwise choice [6, 10].

3.2. PetriNets

A Petri net [9] is a more concise and intuitive way of representing a situation to be modeled. It is also useful to automate the generation of large state spaces. A Petri net consists of places, transitions, arcs and tokens. Tokens move from one place to another along arcs through transitions. The number of tokens in the places represents the marking of a Petrinet. If the transition ring times are stochastically timed, the Petri net is called a Stochastic Petri Net (SPN). If the transition ring times are exponentially distributed, the underlying reachability graph, representing transitions from one marking to another gives the underlying homogeneous CTMC for the situation being modeled.

For the paralell system, we are interested in finding the probability that an incoming task is turned away because all n processors are tied up by other tasks being processed. The parameters associated with this pure performance model are, arrival rate of tasks, service rate of tasks, the number of task response times. The performance model assumes that the arriving task forms a Poisson process of rate and the service requirements of tasks are independent, identically distributed with the exponential distribution. A deadline d is associated with each task. Let us also take the number of buyers available for storing incoming tasks as b. Place process contains the number of processors available. Initially there are n tokens here representing n processors. In general PetriNets are a graphical method to analyze the process, but is not a mathematical method or an algorithm to solve the problems of thorughtput calculation.

3.3. Reliability Block Diagrams

In Reliability Block Diagrams each component of the system is represented as a block [1, 9]. The blocks are then connected in series, parallel, stand-by, and so on, based on the operational dependency between the components. If for the system to be up all the components need to be operational, blocks in a RBD are connected in series. On the other hand, if the system can survive with at least one component then blocks are connected in parallel, this logics is extended to other logics like partial redundancy, stand-by and even buffer systems. An RBD can be used to model availability if the repair times (and failure times) are all independent. Fig 3 shows availability model with n machines where all machines are required for the system to be up. From this we conclude that the RBD represents a simple series system. Given a failure rate λ_i and repair rate μ_i, the availability of each machine is given by:

$$A_i = \left(\frac{\mu_i}{\lambda_i + \mu_i} \right)$$

(5)

So the availability of the system is:

$$A_s = \prod_{i=1}^{4} A_i \qquad (6)$$

Figure 3. Series system.

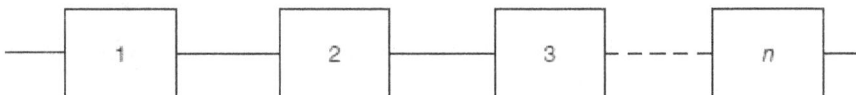

Figure 3 shows a multiprocessor availability model with n processors where at least one processor is required for the system to be up. From this we conclude that the RBD represents a simple parallel system. Given a failure rate λ and repair rate μ, the availability of the system is:

$$A_s = 1 - \prod_{i=1}^{n} \left(1 - \frac{\mu}{\lambda + \mu} \right) \qquad (7)$$

Figure 4. Multiprocessor system.

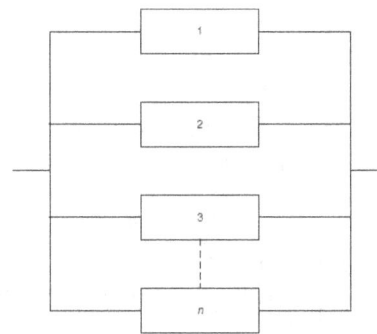

Reliability Block Diagram allows to mix the systems and the n, for example, the multiprocessor system will be now just a block, which could be a part of a series system. Thus the net is reduced to have a single availability for the entire system.

To involve production estimation trough RBDs is only necessary to do a multiplication between the availability of the system with the nominal production rate of the plant and the evaluated time.

The result of the calculus is then an expected total availability or production, which is useful when is needed to do a forecast of the total production in a very long-term analysis.

4. PROPOSAL ABOUT THE PROBLEM

By itself, the simulation corresponds to the process of designing a model in which different input variables, with the purpose to study the output variables generated by the model and evaluate and make decisions. In terms applied to the current context, the simulation corresponds to a non-deterministic method based on the statistics used to analyze complex expressions that could be extensive in terms of time and cost to assess accurately. Simulation allows to study how the items interact, analyze their compatibility, criticity and effects that might occur for which no consideration was.

Figure 5. Basic diagram for representation of the operation of a simulation in the area of operations.

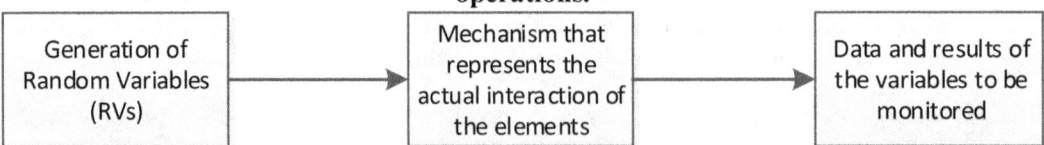

4.1. Reliability and Maintainability Modelling

Considering what reliability engineering theory raises, there are mainly two random processes involved: the maintainability function and the reliability function. How has been mentioned in the state of the art, the reliability, denoted as $R(t)$, corresponds to the probability that an item could work without failures during a time t, since in t=0 it has the condition as good as new or other state, given by the specific modelling method (PRP, NHPP, GRP), and the maintainability $m(t)$ corresponds to the probability to repair an item before a time t, since t=0 is when the repair task was initiated. Both expressions are defined from probability distribution functions, and represents the random variables Time To Fail (TTF) and Time To Repair (TTR), related to reliability and maintainability respectively.

To run a simulation of a system is necessary to take as input the information indicating how the equipment interacts, it means the logic-configuration in which are related. So the input variables are:

1. Behavior of times to fail of each equipment. (Random Variable)
2. Behavior of times to repair of each equipment. (Random Variable)
3. How an item failure or an item repair impacts in the others items. In other words, how the items are related in terms of logic-configuration under which they are interacting. (Deterministic information).

We have to remark that the reliability functions $R(t)$ and maintainability $m(t)$, are defined by a fitting method to the historic data, if there is. The most suitable method is maximizing the likelihood to each distribution. Once the parameters to each distribution that maximizes the likelihood are determined a rank is generated with the Kolmogorov-Smirnov test. The process of choosing the best distribution once the parameteres are estimated it may even visually, but there are also tests like the Anderson-Darling and Shapiro-Wilk, when is more related to normal distributions.

Figure 6. Representation of interactions of an operating system of 3 the items.

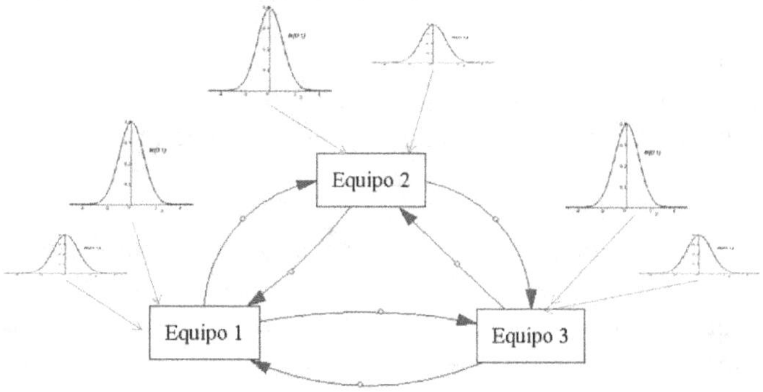

For this methodology have been considered various types of events occurring in the production environment, and certain types of interactions that the equipment have, which corresponds to existing logic-configurations between them.

Within the behavior of the equipment, the items have the following attributes:

- Times between random failures, associated to a certain pdf.
- Random repair times, associated to a certain pdf.
- Time between operational stoppages, associated to a certain pdf.
- Duration of operational stoppages, associated to a certain pdf.
- Some certain productive capacity of the item.

By the side of the subsystems we have considered the following types of logic configurations:

- Series configuration
- Parallel configuration.
- Partial redundancy configuration, aka k-out-of-n

- Stand-by configuration
- Load-Share configuration
- Stockpile, aka buffer

Probability distributions have been considered for use in the model correspond to: Exponential, Log-Normal, Normal, Dirac delta, Uniform and Weibull of 2 and 3 parameters. Also NHPP and GRP modelling is allowed. To the random number generation the inverse distribution method is used.

4.2. Modelling Elements

4.2.1 Equipment Tree
The Equipment Tree corresponds to the hierarchy and detailed component-level constitution of all the equipment that form the system, given through a hierarchical table at which level each subsystem and each component/equipment belongs.
The Equipment Tree list all components in the system. Each component has a unique code, a probability distribution of failure time and a unique probability distribution of repair time. The *components* then correspond to the elements that can fail and could be repaired.

4.2.2 Flowcharts
The methodology aims to, through a practical and intuitive insight, accelerate the process of modeling the most, so that practitioners should only have to identify the operating logic based on the material flow in the system, for this purpose is needed to be developed, after obtained the Equipment Tree, the flow diagram of system to model. The diagrams are composed of two types of elements, *nodes* and *arcs*, the first can be of systems, equipment or changes in flow. The *arcs* are easier however, only indicate where to which the material is transported.

4.2.2.1 Nodes
Within the several types of layout nodes and these can be classified into three:

4.2.2.1.1 Processable Nodes
In this type are the nodes representing systems, subsystems and equipment. Systems and subsystems are used as nodes that exclusive group other diagrams, in contrast the equipment nodes are the last elements in the Equipment-Tree.

a) Equipment: are those elements that fail and are repaired, so are the elements on which the data is loaded into:
- Probability distribution of time between failures
- Probability distribution of time for repairs
- Preventive Maintenance Policy
- Maximum production capacity
- Other optionals

b) Subsystems: Are groups of equipment or other subsystems. Don't have specific probability distribution functions for failures or repair. In contrast, the behavior of the subsystem is given by the interaction of the failures and repairs of the elements that compose the subsystem.

4.1.2.1.2 Node of Logic Configuration
These nodes correspond to the method by which the flow could be divided, cumulated, etc. being able to represent different logical configurations. Through these nodes is possible to establish different types of redundancy subjected to different areas of the system.
Each node has its own logical configuration algorithm, based on state changes in each item (equipment) or subsystem that is defined in your environment flux levels being processed and communicated immediately to the other elements the impact of the last event in itself.

4.2.2.1.3 In and Out Nodes

They are the last type of node, but no less important, since its purpose is to represent where the mass flow is delivered and where it comes from, so these nodes open and close the flow. These are nodes in ideal conditions, since they do not experience failures.

The In and Out nodes are within any system or subsystem diagram that can be decomposed.

4.2.2.1 Arcs

Graphically correspond to arrows, from where to where there is a transfer of mass flow. Indicates the start and destination. Not correspond to elements that may experience failure or a stoppage, only serve to connect the existing mass flow between elements.

4.3 Types nodes of configuration logic
4.3.1 Series

This type of configuration is when the dependency between certain equipment operations requires that to the existence of system availability requires that all equipment is in a state of availability. Any unavailability, independent of which item (equipment) will generate an overall unavailability of the system or subsystem.

The layout of a subsystem in this configuration corresponds to:

Figure 7. Layout for equipment under series configuration.

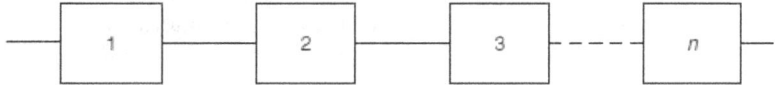

4.3.2 Load-Share

This configuration represents the case where a group of machines (equipment/subsystem) share the mass (flow) passing through a production line. This distribution may or may not be equal to each machine. This depends exclusively on the characteristics of capacity available to each machine.

The benefit of this type of functional logic is that it gives the possibility to operate at lower production rates than the maximum of the line, ie the failure of a machine only generates a proportional or partial loss in production, but not the unavailability entire system. Each machine (equipment/subsystem) has a level of impact (in percentage) on the production line to which it corresponds. Can also be configured the availability in excess of capacity, in this way is the case when the sum of the impacts that the machines have on the line is above 100%.

4.3.3 Partial redundancy and Parallel (k-out-of-n / 1-out-of-n)

The parallel configuration or full redundancy occurs when there is a group or sub-system consisting of N equipment, where the unavailability of one or more components, up to N-1, generates no impact on the overall operation, always requiring either conforming equipment cannot meet the performance of the entire subsystem. Thus, the unavailability of this system is defined if and only if all equipment are unavailable.

Figure 8. Diagram for partial redundancy configuration.

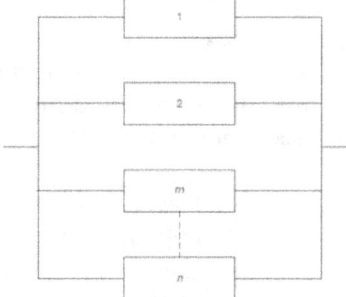

4.3.4 Stand-by

This type of functional dependency can represent the operation of a group of machines where one works at the same time, and if it fails instantly goes into operation other machine which is as a backup. There may be more than one machine as a backup, as outlined in the following diagram:

Figure 9. Diagram for stand-by configuration.

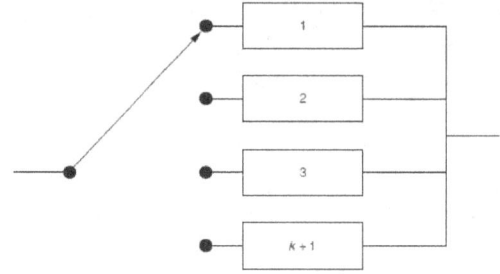

4.3.5 Conditioned Flow

Is also possible to use another configuration, the conditional flow, where a priority is assigned to different machines to be able to deliver the material. This router node sets the order of preference you want to give to a list of equipment or subsystems to processes the material flow.

4.3.6 Stockpile / Buffer

A stockpile is characterized by allowing the accumulation of material in the form of batch or continuous within itself, getting these as a result of a previous process and arranging to be taken as input in a subsequent process. This method is limited by the maximum capacity of the stockpile and dynamic flows that are in the subsystem upstream and downstream.

Figure 10. Representation of accumulator system between 2 processes.

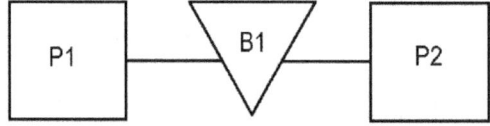

Thus, depending on: inflows, outflows, capacity, and level of accumulation that possess at a given time, the stockpile can be be in different states of accumulation. The fundamental operation of the stockpile is to provide material to the downstream subsystem while the upstream system is in a state of unavailability, thus achieving greater advantage of the available hours of the downstream system, which logically should be part of the bottleneck production system in terms of capacity, since otherwise the need to implement a storage system loses relevance and prominence, since in this way the system upstream would require constant production in their hours of operational readiness given to the system downstream may prosecute higher the output rate of the system upstream. The variables that define the stockpile are: Maximum Capacity [$Q_{máx}$], Initial Capacity [Q_{in}], Downstream Rate [F_{ad}], Upstream Rate [Fu_a].

4.4 Representation and calculation on the system.

The method of processing subsystems aims to identify all the interactions generated by analyzing what happens as a result of each event in the life line of them and finally get a final life-line for each of the subsystems, and this life-line could be used again in case if another subsystem contains the previous one.

Processing subsystems depends purely on dependencies configuration that the elements have, for this purpose various algorithms dedicated to each configuration were prepared. We have to note that each subsystem is linked to a dependencies configuration, and in the row for that subsystem with unique parameters related to the configuration that has been associated , for example in the case of partial redundancy variables are given: how many are required for the subsystem is available, in Load-Share the production rate, etc.

4.4.1 Calculation focused to the state change.

The aim is that the processing methodology and calculation of indicators would be the fastest way possible. Most simulators for continuous processes perform the calculation of indicators and identification of states through monitoring a certain interval of time, this process is extremely inefficient when compared with oriented view state changes. It means that monitoring is performed only when something in the system changes state. For this study, all occurrences of fault level machine are independent therefore a simulation oriented state change.

Figure 11. Outlining the algorithm scanning system in parallel.

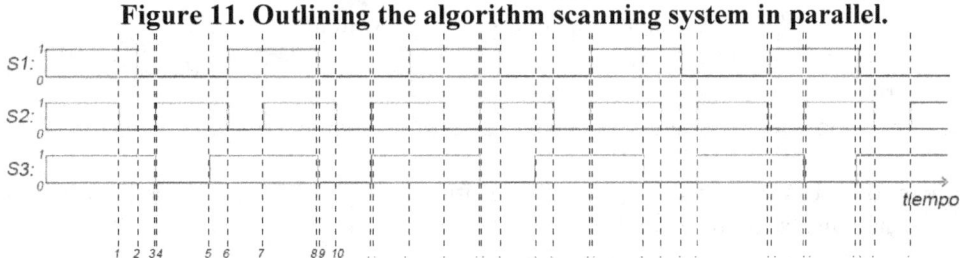

This algorithm works asking the state each time an item will change the state, then the impact of this change of state is analyzed in each of the system elements. Now the way that the event impacts on other elements depends on the established functional logic. Which have been defined in 4.3

These logical relationships are embodied in a multidimensional array R where recursively runs the impact on each element.

$$R = \begin{bmatrix} R_{1,1} & \cdots & R_{1,j} & \cdots & R_{1,n} \\ \vdots & \ddots & & & \vdots \\ R_{i,1} & & R_{i,j} & & R_{i,n} \\ \vdots & & & \ddots & \vdots \\ R_{n,1} & \cdots & R_{n,j} & \cdots & R_{n,n} \end{bmatrix} \quad (8)$$

n: total number of processables nodes
i: id of the element that suffered a change of state.
j: id of the element on which there is an impact on its state because of the change on element i.

At the same time each vector is defined as the following:

$$R_{i,j} = \{v_{i,j}(f_i), d_{i,k}, t_{i,k}\} \quad (9)$$

Where:
$v_{i,j}(f_i)$: instant mass flow on element j given that the mass flow on i now is f_i
$d_{i,k}$: type of stoppage on the k-th stoppage of element i.
$t_{i,k}$: time of stoppage of the k-th stoppage of element i.
$p(d_{i,j,k})$: Binary variable which indicates if the change of state of i generated a stoppage in the element j (ie. Change in mass flow)

Once analyzed the impact and collected the instantly generated variables other algorithm is run, that studies which will be the next event to occur. That could be: Failure of element, End of repairing of an element, Change of mass flow in an element, Empty stockpile, Maximum level of a stockpile has been reached. The algorithm processes and determines what the next event to occur in the system, through the use of a procedure that calculates the remaining life time of each element. The time at which this event will occur is determined and is used again in the matrix R. For each subsystem a life-line is generated, which is saved in a matrix:

$$S_i = \{P_1, P_2, P_3, ..., P_{l_i}\} \quad (10) \text{ ,where:}$$

P: Vector which has the information of mass flow at a given time.
l_i: Total number of change of stated of the element i

$P_x = \{m, t, g\}$ (11), where:

m: mass flow

t: time

g: new state of the element (working, ie. suffering degradation, or in stoppage)

4.4.2 Implementation of the calculation

The presented method can be used to calculate:
- Total number failures, per element, subsystem, and system
- Total number of system detention
- Number of failures dumped by a stockpile system
- Total mass flow for an established simulated period of time

Given the method is possible to use the overall algorithm to analyze the risk based a large number of simulations in short times. In this way, the methodology is useful when is about risk measuring. For example, with the method is possible to calculate with accuracy the probability of meeting a required amount of production for an established period.

5. CONCLUSION

A simple, reliable and fast method to simulate systems focused in the occurrence of failures has been presented. This methodology presents the advantage that is not necessary to define stochastic Petri-Nets states or States of the whole system like in Markov Chain modelling, is only necessary to establish the dependencies, from whom we have considered the most common seem in the industry. The work of establish the state-based models by hand is a task that took long time, and is highly vulnerable to mistakes in the process of establishing these states.

To use the proposed modelling no previous knowledge about RBD, Markov Chain or Petri-Net is required. The requirement to the user is just to identify the functional relationship between the machines/elements. The methodology, as a state change based simulation method, allows to run a large number of simulations, and as a difference with other state-change based simulation environments, the present methodology only focuses only limited states, so the speed is increased dramatically. This allows to study a larger number of cases and a large number of combinations and possibilities.

Lastly, this methodology represent disadvantages in terms of events that are not related to faults and its implications. The method also oriented calculations only consider long term or transient states. The calculations do not consider production states of calibration, but indicated by average production machine. So, further work could be related to fix those approaches.

Acknowledgements

Thanks for RelPro® for allowing the funding for this research.

References

[1]   Introducing buffer inventories in the RBD analysis of process production systems, Marco Macchi, Fredy Kristjanpoller, Marco Garetti, Adolfo Arata, Luca Fumagalli Reliability Engineering & System Safety, Volume 104, August 2012, Pages 84-95

[2]   Reliability and availability analysis of low power portable direct methanol fuel cells N.S. Sisworahardjo, M.S. Alam, G. Aydinli Journal of Power Sources, Volume 177, Issue 2, 1 March 2008, Pages 412-418

[3]   Chapter 4 - Reliability, Availability, and Maintainability Analysis, Eduardo Calixto, Gas and Oil Reliability Engineering, 2013, Pages 169-347

[4]   Make your Petri nets understandable: Reliability block diagrams driven Petri nets, Jean-Pierre Signoret, Yves Dutuit, Pierre-Joseph Cacheux, Cyrille Folleau, Stéphane Collas, Philippe Thomas, Reliability Engineering & System Safety, Volume 113, May 2013, Pages 61-75

[5]   Probabilistic analysis of a series–parallel repairable system with three units and vacation, Linmin Hu, Dequan Yue, Jiandong Li Applied Mathematical Modelling, Volume 34, Issue 10, October 2010, Pages 2711-2721

[6]   A novel optimal preventive maintenance policy for a cold standby system based on semi-Markov theory, Chongquan Zhong, Haibo Jin, European Journal of Operational Research, Volume 232, Issue 2, 16 January 2014, Pages 405-411

[7]   Availability/reliability engineering analysis of three, four, and five-stage unreliable series transfer lines, Thomas F. Hassett, Duane L. Dietrich, Journal of Manufacturing Systems, Volume 14, Issue 6, 1995, Pages 427-438

[8]   A simulation study on maintainer resource utilization of a fast jet aircraft maintenance line under availability contract, Partha Priya Datta, Anupam Srivastava, Rajkumar Roy, Computers in Industry, Volume 64, Issue 5, June 2013, Pages 543-555

[9]   Simulation inferences for an availability system with general repair distribution and imperfect fault coverage, Jau-Chuan Ke, Zheng-Long Su, Kuo-Hsiung Wang, Ying-Lin Hsu ,Simulation Modelling Practice and Theory, Volume 18, Issue 3, March 2010, Pages 338-347

[10]   Fuqua, Norman B., "Markov Analysis," The Journal of the RAC, Third Quarter 2003.

[11]   Bause, F., Kritzinger, P.: Stochastic Petri Nets – An Introduction to the Theory (2nd edition), 2002,

[12]   Xin-Feng He, Su Wu, Quan-Lin Li, Production variability of production lines, International Journal of Production Economics, Volume 107, Issue 1, May 2007, Pages 78-87

[13]   M. Manitz, Tempelmeier H., The variance of inter-departure times of the output of an assembly line with finite buffers, converging flow of material, and general service times. OR Spectrum, January 2012, Volume 34, Issue 1, pp 273-291

[14]   Colledani, M., Matta, A., and Tolio, T. (2010). Analysis of the production variability in multistage manufacturing systems. CIRP Annals - Manufacturing Technology, 59(1), 449-452.

[15]   Congbo Li, Ying Tang, Chengchuan Li, Lingling Li: A Modeling Approach to Analyze Variability of Remanufacturing Process Routing. IEEE T. Automation Science and Engineering 10(1): 86-98 (2013)

[16]   V. Krivtsov, O. Yevkin, Estimation of G-renewal process parameters as an ill-posed inverse problem, Reliability Engineering & System Safety, Volume 115, July 2013, Pages 10-18

[17]   Safety and Reliability Society, SARS, Applied R&M Manual for Defence Systems, Part D - Supporting Theory, GR-77 Issue 2011, 2011.

[18]   Jingshan Li and Semyon M. Meerkov. Production variability in manufacturing systems: Bernoulli reliability case. Annals of Operations Research 93 (2000) 299–324.

[19]   Monika Tanwar, Rajiv N. Rai, Nomesh Bolia, Imperfect repair modeling using Kijima type generalized renewal process, Reliability Engineering & System Safety, Volume 124, April 2014, Pages 24-31

[20]   Hoang Pham, Xuemei Zhang, NHPP software reliability and cost models with testing coverage, European Journal of Operational Research, Volume 145, Issue 2, 1 March 2003, Pages 443-454

[21]   S. Zarezadeh, M. Asadi, N. Balakrishnan, Dynamic network reliability modeling under nonhomogeneous Poisson processes, European Journal of Operational Research, Volume 232, Issue 3, 1 February 2014, Pages 561-571

[22]   Yeu-Shiang Huang, Chi-Chang Chang, A study of defuzzification with experts' knowledge for deteriorating repairable systems, European Journal of Operational Research, Volume 157, Issue 3, 16 September 2004, Pages 658-670

[23]   Marco Scarsini, Moshe Shaked, On the value of an item subject to general repair or maintenance, European Journal of Operational Research, Volume 122, Issue 3, 1 May 2000, Pages 625-637, ISSN 0377-2217

[24]   Tim Bedford, Isha Dewan, Isaac Meilijson, Athena Zitrou, The signal model: A model for competing risks of opportunistic maintenance, European Journal of Operational Research, Volume 214, Issue 3, 1 November 2011, Pages 665-673

Risk Quadruplet: Integrating Assessments Of Threat, Vulnerability, Consequence And Perception For Homeland Security

Kara Norman Hill[a] and Adrian V. Gheorghe[b]
[a] Booz Allen Hamilton, Norfolk, VA, USA
[b] Old Dominion University, Norfolk, VA, USA

Abstract: Risk to a critical infrastructure, is considered to be a function of threat, vulnerability, and consequence. It has long been a challenge to integrate these three disparate assessments to establish an overall picture of risk to a given asset. There are many different types of risk assessments performed on assets and those different assessments explore risk from different perspectives. Is the asset a critical power plant, essential to electricity generation? Is it a large dam, critical to the water supply? Is it a major road, critical to transportation? Or is it a major tourist attraction, critical to national morale? Or, like the Hoover Dam, is it all of these things? Which risk assessment is "right"? How can all of these risk assessments be integrated? Are certain risk assessments more important than others? Obviously, risk is a function of our perceptions, which can influence our understanding of threat, vulnerability, and consequence. A risk quadruplet methodology is proposed to systematically integrate risk perceptions with assessments of threat, vulnerability, and consequence in a traceable, reproducible, and meaningful manner. The risk quadruplet model is explored by leveraging Evidential Reasoning technique (MCDA), along with simulated data for threat, vulnerability, consequence, and perception.

Keywords: Systems Engineering, Risk Management, Critical Infrastructures, Perception, Evidential Reasoning

1. INTRODUCTION

Many talk about risk as a function of threat, vulnerability, and consequence [4, 9]. Multiple risk assessments, which seek to assess threat, vulnerability, and consequence to a specific asset or facility, could vary widely [6]. Risk assessments could be based on risk data or perceptions. The data from one assessment could be drastically different from the data of another assessment; one assessment could incorporate factors such as whether the risk was voluntary or involuntary, while another might attempt to calculate risk using traditional risk equations [8].

Figure 1. Proposed Risk Quadruplet ©.

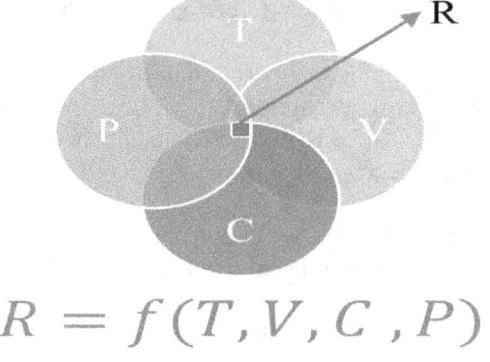

$$R = f(T, V, C, P)$$

There is also confusion about the definitions of threat, vulnerability, and consequence, let alone how to assess those nebulous concepts. The many different definitions of these concepts can drastically affect risk calculations. Threat could be viewed as a single scenario, or the likelihood of that scenario. Vulnerability could be seen as a probability, or it could be viewed as a state of the system, from which conditional probabilities of threat might be derived. And there are many types of consequences (economic, environmental, or in some cases loss of life), which must all be assessed in order to give the best possible overall risk picture. Most of this confusion arises from our inherent perceptions. There is, inevitably, an element of subjectivity to any risk assessment, and that subjectivity is currently missing from the risk assessment approach. It only makes sense to integrate our T, V, and C assessments with our perceptions into an overall, improved, risk assessment approach, thus defining a new risk paradigm. A risk quadruplet is proposed in this dissertation that incorporates threat (T), vulnerability (V), consequence (C), and perception (P) (Figure 1).

We must first define what we are trying to protect, which is the collection of critical infrastructure (CI), key resources (KR), and key assets (KA). Then we must define the way in which we shall protect those CIKRKA, which is to determine their overall risk as a function of T, V, C, and P, then rank them accordingly, such that risk mitigation actions can be prioritized and implemented. Figures 2-3 provide a list of terms and their intended meanings when used throughout this paper. Some of these definitions are pulled straight from the literature. Others are modified from definitions provided in official, government documents, such as the Department of Homeland Security (DHS) Risk Lexicon [7]. All of these definitions, as they are presented here, reflect the intents and purposes of this research.

Figure 2. CIKRKA Definitions.

CI	Government and private systems essential to the operation of our nation in any or all aspects of the lives of its citizens (health, safety, economy, etc.), such as utilities, facilities, pipelines, etc.
KR	Public or private resources essential to the operation of our nation's government and economy, such as fuel or goods.
KA	Those buildings, geographic regions, monuments, or icons, whose destruction would cause a crushing blow to our nation's ego, morale, and identity, but which are not essential to the operation of our nation, such as the Washington Monument or the Statue of Liberty.

The belief is that the currently accepted homeland security risk triplet (T, V, and C) is inadequate for characterizing risk to CIKRKA and that a risk quadruplet should be explored to incorporate perception into the current risk assessment approach. But exactly how those components of risk are integrated must be decided. The improved risk assessment integration methodology, based on T, V, C, and P assessments, will be developed and presented. This methodology will systematically integrate all four assessments in a meaningful, traceable, and reproducible approach using systems engineering techniques such as risk analysis and Multi Criteria Decision Analysis (MCDA). The end result will be a ranking of CIKRKA, based on the risk quadruplet methodology. This will allow for a more comprehensive ranking of these disparate entities along multiple risk scales. This ranking system will improve resource allocation for risk mitigation efforts in support of homeland security and homeland defence missions.

Figure 3. Risk Definitions.

Threat	Threat of a risk scenario to an asset; threat of an intentional risk scenario is generally estimated as the likelihood of an attack (that accounts for both the intent and capability of the adversary) being attempted by an adversary; for other risk scenarios, threat is estimated as the likelihood that the risk scenario will manifest; however, threat can also be estimated qualitatively as perceived likelihood
Vulnerability	Ability of an asset to endure a risk scenario despite physical features, operational attributes, characteristics of design, location, security posture, operation, or any combination thereof that renders an asset open to exploitation or susceptible to a given risk scenario; can be estimated qualitatively, or quantitatively, as the likelihood of a successful risk scenario given the risk scenario is identified, which implies that vulnerability is also related to resilience
Consequence	Effect of a successful risk scenario on an asset; consequence is commonly assessed along four factors: human, economic, mission, and psychological, but may also include other factors such as impact on the environment; consequence can be measured quantitatively if data exists, but can also be measured qualitatively either along a set of scales or along a single integrated consequence scale for which all consequence factors are considered as a whole

Perception	Subjective judgment about the severity of a risk scenario to an asset; may be driven by sense, emotion, or personal experience; generally measured qualitatively
Risk	Potential for an unwanted outcome resulting from a risk scenario, as determined by the T, V, C, and P of that risk scenario to an asset; often measured and used to compare different future situations, as well as to rank assets for the purposes of risk mitigation and budgeting for emergency preparedness, response, and recovery

2. RISK QUADRUPLET

Model. The risk quadruplet consists of three phases (Figure 4). The first phase is the perception assessment. The second phase consists of T, V, and C assessments. The final phase is the assessment integration phase, where the assessments of T, V, C, and P are all assimilated.

Figure 4. Risk Quadruplet Phases ©.   **Figure 5.** Risk Quadruplet Model.

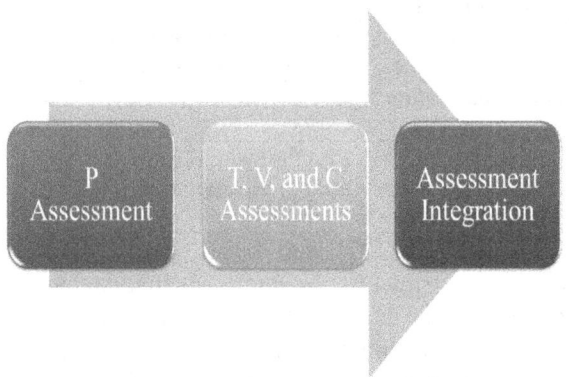
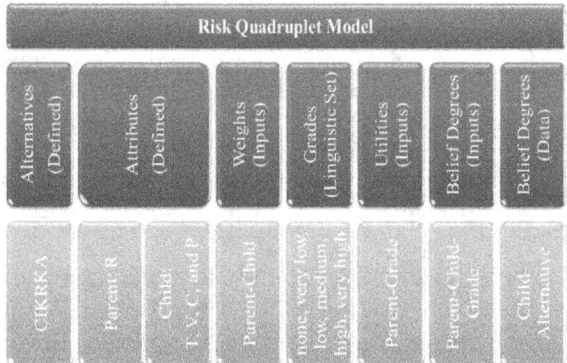

The risk quadruplet model to integrate the perception, T, V, and C assessments is given in Figure 5. It consists of alternatives, attributes, weights, grades, utilities, and belief degrees. The alternatives, in our case, are a set of CIKRKA assets. Further defining the model, we have a parent attribute denoted as risk (the overall value we are seeking to calculate), as well as child attributes (T, V, C, and P), all of which are part of the risk function. We also define grades for the child attributes, as they relate to the alternatives, using a linguistic set (none, very low, low, medium, high, very high). Weights are chosen to relate the child attributes to the parent attribute. Utilities are assigned to relate the grades to the parent attribute. The first set of belief degrees relates grades to the parent and child attributes. In other words, does the linguistic set choice of none for T, V, C, and P directly correlate to a linguistic set choice of none for the parent attribute of risk? What about the choice of very low? If so, the belief degrees assigned to relate those relationships would be higher than those relating a grade of none for a child attribute to a grade of high for the parent attribute.

The second set of belief degrees are derived from the assessment data and are used to relate grades to the alternatives within each child attribute. For the perception assessment, the belief degrees are the proportions calculated based on how many respondents selected each of the linguistic set choices in our adapted psychometric survey to collect risk perception data. For the T, V, and C assessments, the belief degrees would be translated to the linguistic set if the data was leveraged from historical assessments, or that data could be collected in a new set of assessments using the linguistic set.

Methodology. With the model defined, the next problem was how to test its viability. It would be ideal to validate the risk quadruplet methodology in vivo or in the real world, using real data, collected anew, with a full scale model of multiple CIKRKA to compare and rank. However, due to the constraints of scope, cost, and schedule, this type of model Verification, Validation, and Accreditation

(VV&A) is beyond the scope of our research. Instead, we intend to explore this model in vitro, literally in a petri dish, although in our case, the petri dish is a computer (Figure 6).

Figure 6. Risk Quadruplet Viability Testing Options: In Vivo versus In Vitro.

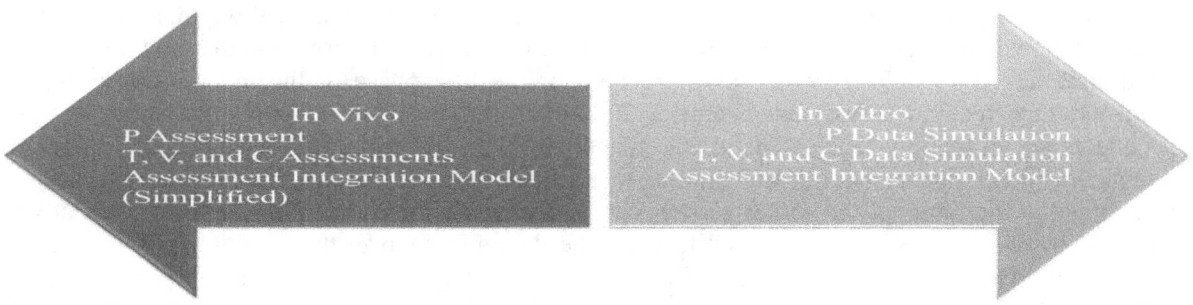

The proposed in vivo risk quadruplet methodology (Figure 7) would consist of the same three phases (assessment of perception; assessments of T, V, and C; and assessment integration) as previously defined (Figure 4). However, we have included additional details on the approach for deploying this methodology. The first phase would consist of a simplified psychometric survey, which would be deployed with a small group of subject matter experts and stakeholders. In order to conduct this survey, we chose Inquisite, a software package capable of deploying surveys online and collecting data [2]. We then designed a questionnaire, choosing a region, risk scenario, and a selection of CIKRKA assets to scope the survey. We also decided to limit the survey (and thus the overall in vivo model) to three CIKRKA assets, and we chose an example for each of the assets. Additionally, to further scope the survey and model, we selected a single risk scenario. We also chose a linguistic set for the survey responses (none, very low, low, medium, high, very high), which would be consistent with the Evidential Reasoning (ER) model we developed.

Figure 7. Risk Quadruplet Methodology (In Vivo).

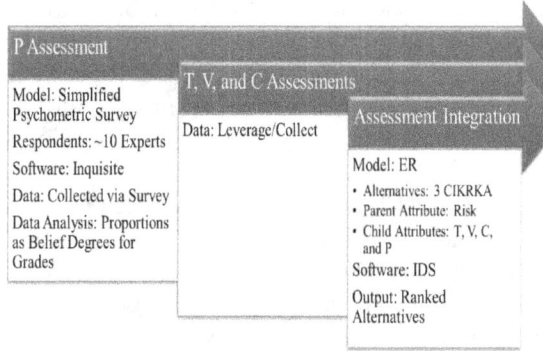

Figure 8. Risk Quadruplet Methodology (In Vitro).

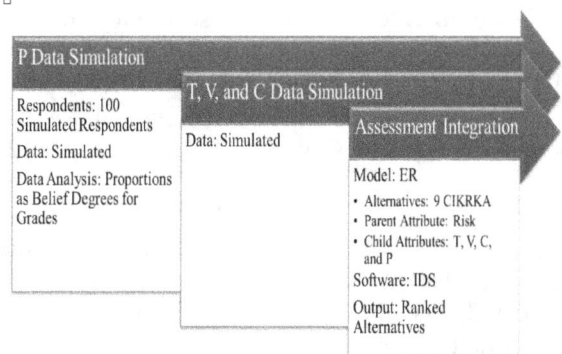

The second phase assumes that the data for T, V, and C could be leveraged from previous assessments, or that those assessments could be conducted. The goal of the risk quadruplet is not to determine how to conduct these assessments, as they are already being conducted and many approaches already exist for doing so, such as the Infrastructure Vulnerability Assessment Model [1]. Rather, the point of the risk quadruplet is to determine how to integrate these assessments with the perception assessment we proposed for the first phase of the methodology. The final phase of the in vivo risk quadruplet methodology focuses on integrating these assessments. The ER model is defined with the three alternatives (CIKRKA assets) used in the Inquisite survey. The parent attribute and child attributes, weights, utilities, and belief degrees are also defined. And the final belief degrees would be input into the model based on the data collected from the perception, T, V, and C assessments. Finally,

Intelligent Decision System (IDS) was the software selected for implementing the ER model, so the alternatives, attributes, weights, utilities, and belief degrees would be input into IDS for analysis [3].

The in vitro methodology is crafted to parallel the in vivo methodology (Figure 8) consisting of the same three phases as the in vivo risk quadruplet methodology; however, there are some obvious differences. The in vitro approach will rely on simulated data to emulate the real world, allowing us to explore the model without risking the exposure of sensitive (in vivo) information that might otherwise jeopardize the very CIKRKA we seek to protect. For the first phase of the in vitro approach, we simulated the perception assessment data using 100 virtual respondents. Rather than rely on leveraging or collecting data for T, V, and C data in the second phase, we simulated this data, as well. Lastly, the assessment integration phase remains similar to the in vivo approach. However, since we are not constrained to the limits of the survey respondents, we increase the number of CIKRKA alternatives. The same software, IDS, would be used to input the data and analyze the results [5]. The resulting analysis would provide a ranked output of CIKRKA assets (alternatives) based on their parent attribute scores (risk). The generalized risk quadruplet methodology (whether in vivo or in vitro) is given in Figure 9.

Figure 9. Risk Quadruplet Methodology.   **Figure 10.** Risk Quadruplet Model (In Vitro).

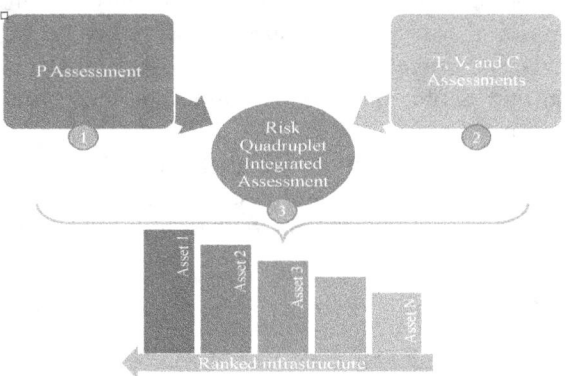 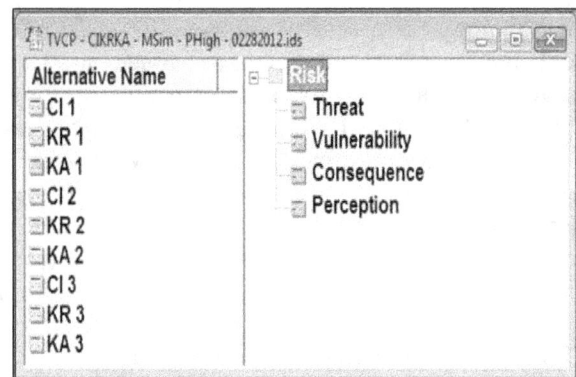

Risk Quadruplet Viability Testing (In Vitro). The in vitro approach for testing the viability of the risk quadruplet methodology relies on simulated data. However, this research is still informative and allows us to explore how the model behaves prior to an in vivo deployment of the methodology. With IDS we are able to build an ER model for the risk quadruplet using a combination of collected perception data and simulated T, V, and C data [2, 3].

An example of how this model appears in IDS is shown in Figure 10. Each of the attributes (T, V, C, and P) were defined and graded using the same linguistic scale. Utilities for the overall or parent attribute (risk) were assigned to these grades (from our linguistic set of none, very low, low, medium, high, and very high) as shown in Figure 11. For our purposes, a risk grade of none would be ideal and thus would receive a Utility of 1. The remaining grades were ranked accordingly. Utilities, unlike probabilities, need not sum to 1.

Figure 11. Grades and Utilities

Grade	Utility [0,1]
None	1
Very Low	.9
Low	.7
Medium	.5
High	.3
Very High	.1

In the interest of keeping this model simple, belief degrees to relate parent and child attributes were assigned using the identity matrix (Figure 12). These belief degrees are not the same belief degrees that are selected by respondents during data collection when they chose the grade they deem appropriate for a given combination of alternative and attribute (collected using the simplified psychometric survey).

Figure 12. Belief Degrees for Relating Parent and Child Grades

Parent Grade/ Child Grade	None	Very Low	Low	Medium	High	Very High
None	1	0	0	0	0	0
Very Low	0	1	0	0	0	0
Low	0	0	1	0	0	0
Medium	0	0	0	1	0	0
High	0	0	0	0	1	0
Very High	0	0	0	0	0	1

Weights are then used to relate the child attributes to the parent attribute. This can be done using visual scoring or using a pairwise comparison of attributes. For the in vitro viability testing, we used the visual scoring approach. Visual scoring is an ad hoc approach, which allows us to visually compare the weights of the different attributes against each other. IDS, initially presents the attribute weights as equal across all attributes. Perception might not be considered equally important by the stakeholders, as the other attributes. We adjusted the weights to create a low perception version of the model for which the perception attribute weight was set to be approximately half as important as the other attributes (where the other attributes were weighted equally and the sum of the weights were constrained to sum to 1) as shown in Figure 13. Other versions of the model will be explored later.

Figure 13. Attribute Weights Using Visual Scoring Attribute (Low Perception)

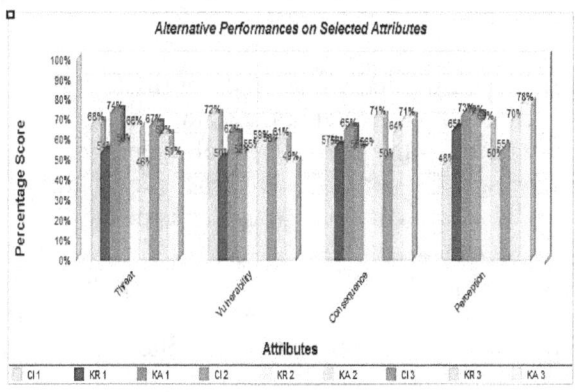

Figure 14. Ranking of Alternatives on Risk (Low Perception)

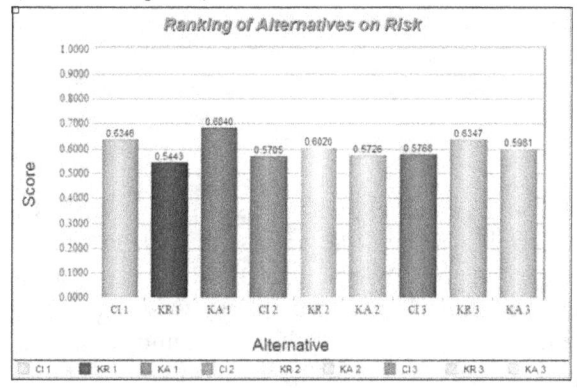

Figure 15. Alternative Performances Across Child Attributes (Low Perception)

Figure 16. KR 1 Grades for Risk Attribute (Low Perception)

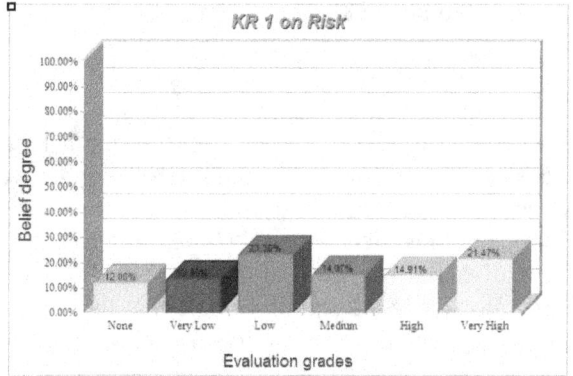

Using simulated data for T, V, C, and P, the IDS model can now rank the nine alternatives (CIKRKA) based on the attributes, grades, and associated utilities, belief degrees, and weights. Figure 14 shows a comparison of the nine CIKRKA alternatives based on their respective overall risk scores. But Figure 15 shows this comparison broken down by the attributes of risk (T, V, C, and P).

Figure 16 shows the breakdown of grades for KR 1 (with the lowest overall risk in the model for which perception was weighted lower than the other attributes) at the parent attribute level (risk). This gives an overall distribution of the calculated grades and belief degrees for risk, based on the grades and belief degrees for the child attributes (T, V, C, and P). Charts can be created by respondents to explore the, degree of belief, indicator.

Model Validation. A preliminary validation of the assessment integration model selected for the risk quadruplet was conducted to determine the impact of selected values, such as weights, utilities, and belief degrees on the ER model. Validation ensures that the model is useful [3]. In other words, the model should address the correct problem and provide accurate information about the system or phenomenon being modeled. Validation of complex models involves demonstrating that the model has the appropriate underlying relationships to permit an acceptable representation of the real world, often exploring the range of inputs under which the risk quadruplet model results are useful.

Figure 17. Risk and Attributes Radar Plots by Alternative (Low Perception)

Figure 18. Risk and Perception Trade-Off Analysis (Low Perception)

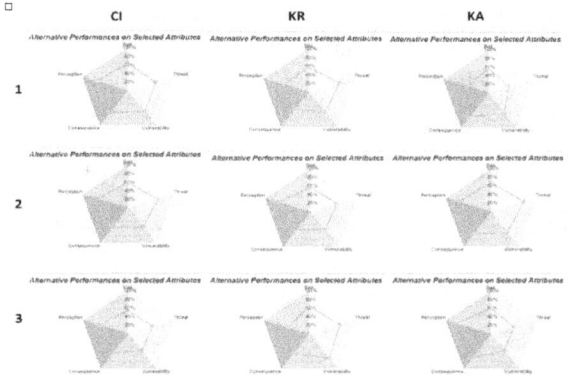

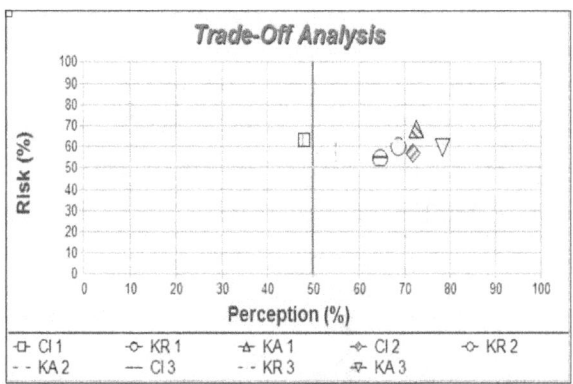

IDS offers some built-in sensitivity analyses and even though this data is simulated, it is still interesting to explore the results as it is obvious how they could be invaluable to the in vivo risk quadruplet methodology [2, 3]. The radar plot shows the values of all of the child attributes, alongside the parent attribute, so it is easy to see which of the child attributes is driving the overall risk score (Figure 17). We can see, for example, that consequence shows some influence on KA 1, while perception affects KR 2 for the low perception model. Figure 18 displays a trade-off analysis chart, which shows the overall risk scores for the nine CIKRKA alternatives, as well as the perceived scores for the low perception model. We see that the overall risk score for KA 3 was 60% even though it was perceived to be 78%, whereas the overall risk score for CI 1 was approximately 63% while it was only perceived to be 48%.

More formally, IDS can produce sensitivity analyses based on the individual child attributes [2, 3]. The graphic given in Figure 19 displays the overall risk scores for each alternative as the weight of the perception attribute is varied from 0 through 1 (we adjusted the y-axis scale, used for the overall risk score, in order to better see the relationship between the weight for perception and the risk rankings). Since we conducted this sensitivity analysis from the high perception model, that value is displayed as a vertical line, denoted as "Given weight", on the chart so that users can compare their current alternative risk scores and rankings to those that would be produced by adjusting the weight for perception. It is interesting to note that the overall risk score for each asset varies with the weight of the perception attribute, but it is not a linear relationship. And while the majority of the alternative risk scores increase as the weight of perception increases, three of the assets show a negative correlation (CI 1, KA 2, and CI 3).

Figure 19. Sensitivity Analysis of Perception

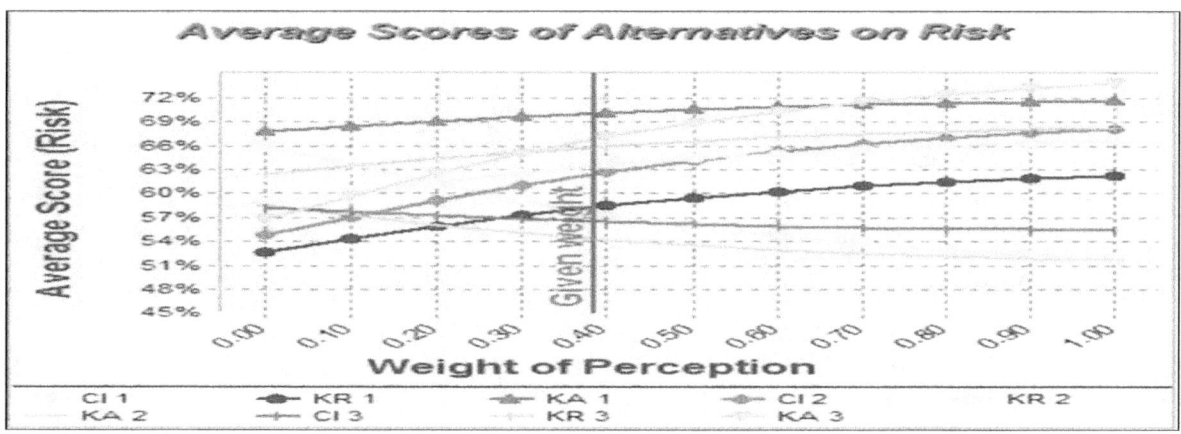

IDS can also produce sensitivity analyses of belief degrees based on adjusting the child attribute weights [2, 3]. We explored only two alternatives from the high perception model: CI 1 and KA 3, ranked lowest and highest on their overall risk scores, respectively (Figure 20). This shows the belief degrees (our simulated data) for the perception attribute related to the grades (our linguistic set) based on the weights input for the child attributes of T, V, C, and P. However, even as we adjust the child attribute weights, the belief degrees do not change, and with good reason. If we recall the belief degree values we chose for relating child attributes to parent attributes (Figure 15), we used the identity matrix; the belief degrees input from our simulated data for the perception attribute would not be impacted by adjusting the child attribute weights. IDS can also produce sensitivity analyses based on the data, itself (Figure 22). The first graph displays the belief degrees input for each grade (from our simulated data) for a selected alternative. We selected KA 3, which received the highest perception score (in the model for which perception received a higher weight). The second graph displays the perception score for all of the alternatives (other attributes, such as T, V, and C can also be explored as desired). Although we did not drastically alter the belief degrees from the original values, we still see a marked change in the overall perception score for KA 3, which dropped from 78% to 68% (Figure 23).

Figure 20. Child Attributes on Belief Degrees (Original)

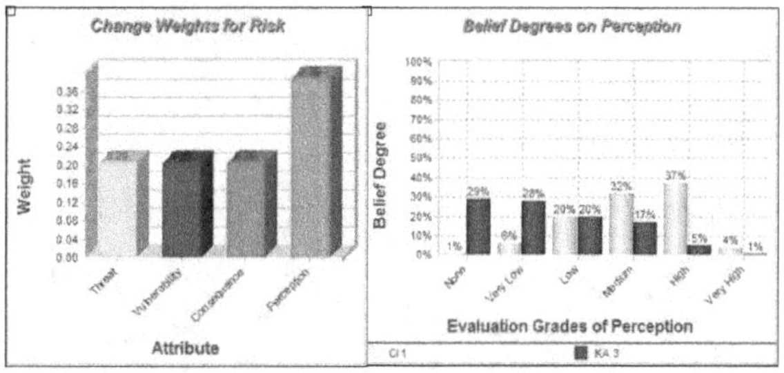

Figure 21. Child Attributes on Belief Degrees (Adjusted)

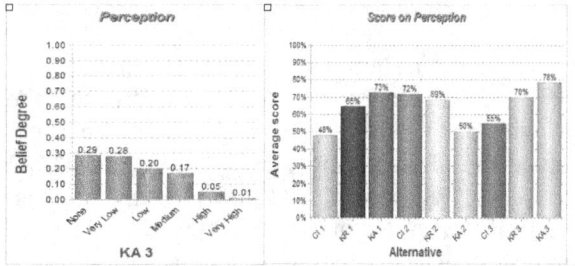

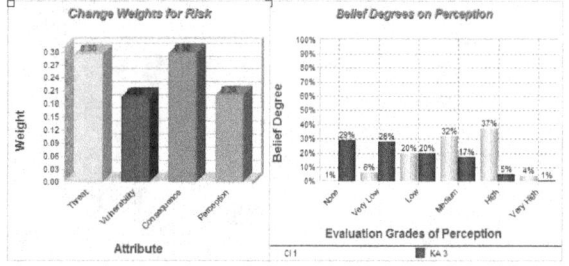

Figure 22. Input Data (Original)

Figure 23. Input Data (Adjusted)

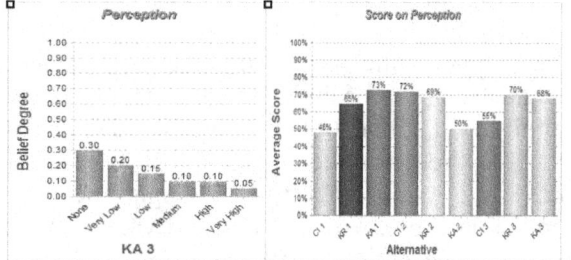

Figure 24. Max Attribute Models

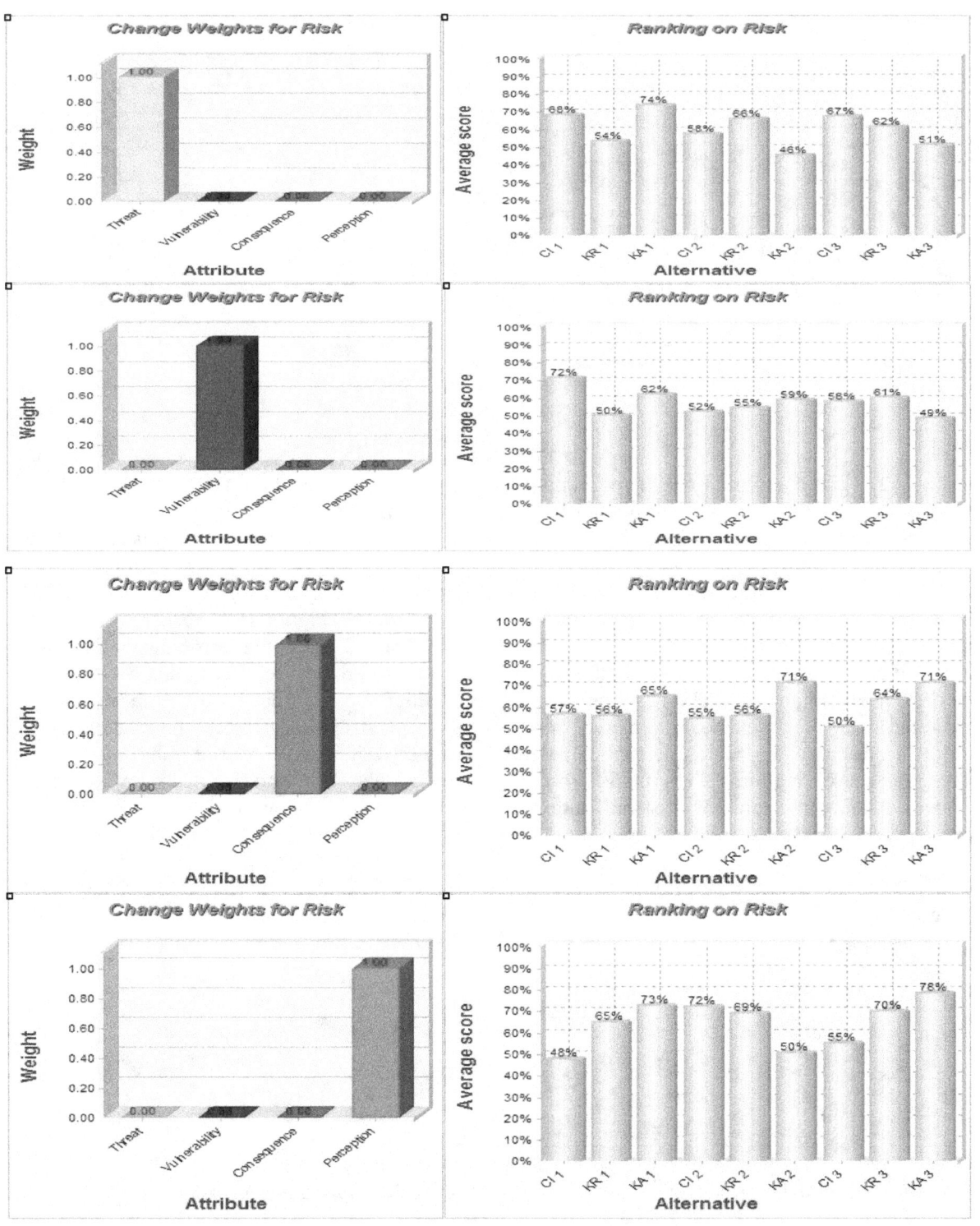

The output of the risk quadruplet model (the ranked CIKRKA) should change depending on the weights selected for the child attributes, so we will explore some extreme weighting cases to test the validity of the model by ensuring that the results align with our intuitions. From Figure15, we know that KA 1 received the highest threat score, CI 1 received the highest vulnerability score, and KA 2 and KA 3 jointly received the highest consequence score, whereas KA 3 received the highest perception score. We will now systematically explore four max attribute - weighting schemes (for example, the Max Threat Model has a threat weight of 1, with vulnerability, consequence, and perception weights of 0 (Figure 24). We condensed the results of these different models in Exhibit 25 and the highlighted values were the assets, which received the highest overall risk score for that model. For the Max T Model, we would expect KA 1 to be ranked highest as it received the highest threat score, and that is exactly what we see. Since CI 1 received the highest vulnerability score, we expect to see it ranked the highest for risk in the Max V Model and that is again what we see. KA 2 and KA 3 jointly received the highest consequence score, so it is no surprise that we see both of them tied for the overall risk score in the Max C Model. And because KA 3 received the highest perception score, it only makes sense that KA 3 received the highest overall risk score for the Max P Model [10].

Figure 25. Model Validation Comparison of Weighting Schemes

	Risk (Max T)	Risk (Max V)	Risk (Max C)	Risk (Max P)
CI 1	68%	72%	57%	48%
KR 1	54%	50%	56%	65%
KA 1	74%	62%	65%	73%
CI 2	58%	52%	55%	72%
KR 2	66%	55%	56%	69%
KA 2	46%	59%	71%	50%
CI 3	67%	58%	50%	55%
KR 3	62%	61%	64%	70%
KA 3	51%	49%	71%	78%

3. CONCLUSIONS

This research challenges the existing paradigm for risk, not just as it is defined in homeland security (as a function of T, V, and C) but as it is typically defined in risk analysis, in general (as a function of probability and consequence). We assert that risk is inherently related to our perceptions and that we construct risk methodologies and models based on those perceptions. The risk quadruplet methodology proposed is capable of integrating T, V, C, and P assessments. While the risk quadruplet methodology was not deployed in vivo, it has been subjected to preliminary testing and analysis, in vitro, and has proven to be a viable approach for ranking CIKRKA in order to improve decision making for homeland security and homeland defense.

References

[1]   Ezell, B. C. (2007). Infrastructure Vulnerability Assessment Model (I-VAM), Risk Analysis, 27(3), 571-583.
[2]   IDS Multicriteria Assessor Quick Guide (2010), www.eids.co.uk
Inquisite, (2011), from http://www.inquisite.com/
[3]   Intelligent Decision System (2010), http://www.e-ids.co.uk/
[4]   National   Infrastructure   Protection   Plan   (2009), http://www.dhs.gov/xlibrary/assets/NIPP_Plan.pdf
[5]   Norman Hill, K., & Ezell, B. C. (2011) - A Risk Quadruplet for Homeland Security. Paper presented at the INFORMS Annual Meeting, Charlotte, NC.
[6]   Norman Hill, Kara, Gheorghe, A. V., (2012). Old Dominion University, Dept. of Engineering Management and Systems Engineering Risk Quadruplet: Integrating Assessments of Threat, Vulnerability, Consequence, and Perception for Homeland Security and Homeland Defense. Thesis (PhD), Old Dominion University.
[7]   Risk   Steering   Committee: DHS   Risk   Lexicon   (2010),   Retrieved   from http://www.dhs.gov/xlibrary/assets/dhs-risk-lexicon-2010.pdf
[8]   Turner, B. A. (1994), The Future for Risk Research, Journal of Contingencies and Crisis Management, 2(3), 146-156. doi: 10.1111/j.1468-5973.1994.tb00037.x
[9]   Willis, H. H. (2007), Guiding Resource Allocations Based on Terrorism Risk. Risk Analysis, 27(3), 597-606. doi: 10.1111/j.1539-6924.2007.00909.x

[10] Macal, C. M. (2005), Model Verification and Validation. Paper presented at the Threat Anticipation: Social Science Methods and Models, The University of Chicago and Argonne National Laboratory.

Author Biographies

Kara Norman Hill, Ph.D., graduated with her doctorate in Engineering Management from Old Dominion University. She also holds an MS in Statistics from The George Washington University and a BS in Mathematics from Virginia Commonwealth University. She is currently a consultant with Booz Allen Hamilton in Norfolk, VA. Her research interests include risk analysis, risk perception, infrastructure protection, systems engineering, operations research, statistics, homeland security, homeland defense, emergency planning, preparedness, response, and recovery, as well as threat, vulnerability, consequence, and perception assessment.

Adrian V. Gheorghe, Ph.D., received his MSc in Electrical Engineering from the Faculty of Power Engineering of the Bucharest Polytechnic Institute in 1968, his PhD in Systems Science/Systems Engineering from the City University, London in 1975, his MBA from the Academy of Economic Studies, Bucharest in 1985 and his MSc in Engineering Economics from the Bucharest Polytechnic Institute. For many years, he has held a permanent position with the International Atomic Energy Agency in Vienna and he was a Senior Scientist with the Swiss Federal Institute of Technology in Zurich, Switzerland. Currently, he holds the Batten Endowed Chair of Systems Engineering from the Old Dominion University in Norfolk, VA, USA.

Revision of the AESJ Standard for Seismic Probabilistic Risk Assessment (1): Extension and enhancement of accident scenario

Yoshiyuki Narumiya[a], Mitsumasa Hirano[b], Tsuyoshi Takada[c] and Kentaro Hayashi[a]

[a] The Kansai Electric Power Co., Inc., Osaka, Japan
[b] Tokyo City University, Tokyo, Japan
[c] The University of Tokyo, Tokyo, Japan

Abstract: This session consists of a four-part presentation on the amendment of the Standard for Procedures of Seismic PRA for NPPs and introduces significant additions/updates in three chapters, Seismic Hazard Evaluation, Building and Component Fragility Evaluation, and Accident Sequence Evaluation. This presentation introduces the purpose, background, and discussed points of the amendment, e.g. extending scope of application to seismic induced events. Upon the revising the previous standard, we updated various requirements in view of advancements in PRA techniques based on new technological findings after the publication of the 2007 version standard and to improve the quality and transparency of this standard. In particular, the amendment reflects the lessons learned and new findings from Fukushima Dai-ichi accident (the 1F accident) as much as possible: e.g. events caused by earthquake, combined seismic and tsunami events, accident management measures, impact to fuel in spent fuel pool, multi-reactor effects, impact of aftershocks, and impact of land sliding.

Keywords: Seismic PRA, Implementation Standard, Fukushima Accident, Seismic-induced complex event, Accident Scenario

1. INTRODUCTION

Japan has been continuously carrying out research on earthquakes from the first, because of the fact that Japan is one of the countries frequently hit by earthquakes and a world-leading seismically active country. In the earthquake-resistant designs of nuclear power plants, research findings related to earthquakes have been reflected and improvements to analytical evaluation techniques have actively continued. With regard to the seismic PRA as well, the development of its methodology has been advanced in research institutions and some industries. The Standards Committee (SC) of the Atomic Energy Society of Japan (AESJ) developed a standard for Procedure of Seismic PRA for nuclear power plants in 2007, with due consideration to the importance and usefulness of a seismic PRA methodology, through the discussion at the seismic PRA subcommittee under the Risk Technical Committee (hereafter called "RTC") of the SC. The standard specifies the requirement which should have the PRA regarding incidents resulting from earthquake as the initiating events at nuclear power plants during power operation, and the concrete method of filling it as an enforcement standard based on the PRA procedure.

2. Process of Revision

The previous standard published in 2007. At the time, the Nuclear Safety Commission (then) revised "the Regulatory Guide for Reviewing Seismic Design of Nuclear Power Reactor Facilities" and required a seismic hazard analysis and a seismic PRA. The RTC (the Power Reactor Technical Committee at that time) decided to prepare the draft of a seismic PRA standard ahead of other countries and started in July 2004. The RTC established three working groups, the seismic hazard evaluation Working Group (WG), the building and component fragility evaluation WG, the accident sequence evaluation WG, under the Seismic PRA Subcommittee (S-PRA SC) at that time.
The RTC reopened the S-PRA SC and three WGs and resumed the discussion of the Seismic PRA Standard at the timing of regular revision and updating requirements based in the 2011 off the Pacific coast of Tohoku Earthquake. The organizational chart is the Table 1 below.

Table 1: The Organizational Chart of Seismic PRA Subcommittee

Sub Committee	Working Group	Roles and Chapter
The Seismic PRA Sub Committee		• Direction and summarization of conclusions of three WGs • Common chapters (Foreword, Scope of Application, Definition of Technical Terms, Normative references, Evaluation Process) • Collection information and analysis of accident scenarios • Documentation
	the Seismic Hazard Evaluation WG	• Collection of information related to seismic hazard evaluations • the methods for seismic hazard evaluation • the methods for developing seismic ground motions from the seismic hazard evaluation results for use with fragility evaluation
	the Building and Component Fragility Evaluation WG	• Establishing Targets of Evaluation and Damage Modes • Selection of Evaluation Method • Actual Fragility Evaluation • Evaluation of Actual Response • Fragility evaluation
	the Accident Sequence Evaluation WG	• Establishing the Initiating Event • Simulation of the Accident Sequence • Simulation of the Systems • Quantification of the Accident Sequence • Analysis of the Containment Vessel Function Loss Scenario

3. Outline of Updated Points

The revised Seismic PRA Standard can provide risk information to improve safety level of nuclear power plant. The standard includes the important and useful points. Many requirements were updated in view of advancements in Seismic PRA techniques based on new technological findings after the publication of the 2007 standard and the quality and transparency of this standard were improved. Some updated points are as follows:

1) The lessons learned and new findings from severe accidents of Fukushima Dai-ichi nuclear power plants, which were occurred on March 11 of 2011
2) Expansion of the standard scope to complicated events, e.g. an fire caused by earthquake PRA
3) Adding a lots of Appendix (Reference), which are references related to issues that can't be required in the standard because of immature methods

4. Scope of Application

The 2007 standard focused on the earthquake-related accident sequences that lead to serious core damage. The internal fire, internal flood, and tsunami related events that may occur as a result of earthquake were excluded from the scope of the 2007 standard. At that time, the RATC was going to develop a fire PRA standard and an internal flood PRA standard after the seismic PRA standard.

However several complicated events occurred in the 1F accident and the RATC needed to expand the scope of the seismic PRA standard to events caused by earthquake and events related to SFP.

1) SFP

In the 2007 standard, the most important event was core damage, but the 1F accident made it clear that there was a possibility that nuclear fuels in SFP were damaged. In this revised standard, fuel damage sequences in SFP are added to core damage sequences.

2) Events caused by earthquake

The 1F accident showed it important that earthquake occurred various complicated events at the same time including fire or internal flooding event. However requirements of events caused by earthquake PRA are mutually related. The revised standard takes partial charge of full requirements of those PRA and assumes the responsibility of providing seismic hazard evaluation and fragility evaluation of special equipments in earthquake induced events PRA. It is possible to implement those complicated PRAs using both the seismic PRA standard and each external PRA standard.

The relation between external PRA standards is Table 2 below. In case of complicated PRA, it is possible to select adequate requirements and combine them to implement PRA. For example, in case of a fire caused by earthquake PRA, users should select requirements according to guidance sentences "if you evaluate internal fire caused by earthquake," which are marked "FireE" in Table 2.

Table 2 The relation among contents of external event PRA

	Collection Information & Accident Scenario	Hazard Evaluation	Fragility Evaluation	Accident Sequence Evaluation
Seismic PRA Standard	FireE FloodE	FireE FloodE (seismic hazard analysis)	FireE(seismic fragility of SSCs related to internal fire PRA) FloodE(seismic fragility of SSCs related to internal flooding PRA)	
Internal Fire PRA Standard	FireE	(Occurrence frequency of internal fire)	FireE (internal fire fragility of SSCs related to internal fire PRA)	FireE
Internal Flooding PRA Standard	FloodE	(Occurrence frequency of internal flooding)	FloodE (internal flooding fragility of SSCs related to internal flooding PRA)	FloodE

Note: FireE : Internal fire caused by earthquake, FloodE : Internal flooding caused by earthquake
An underlined part : the PRA standard mainly provides the information

5. Evaluation Process

The outline of seismic PRA process is the Figure 1 as below, and is almost same as the 2007 standard. Differences between the 2007 standard and the revised standard are as given below;
- Upgrade of requirements for site & plant walk down
- Additional information; lessons learned from the 2011 off the Pacific coast of Tohoku Earthquake
- Upgrade of accident scenarios

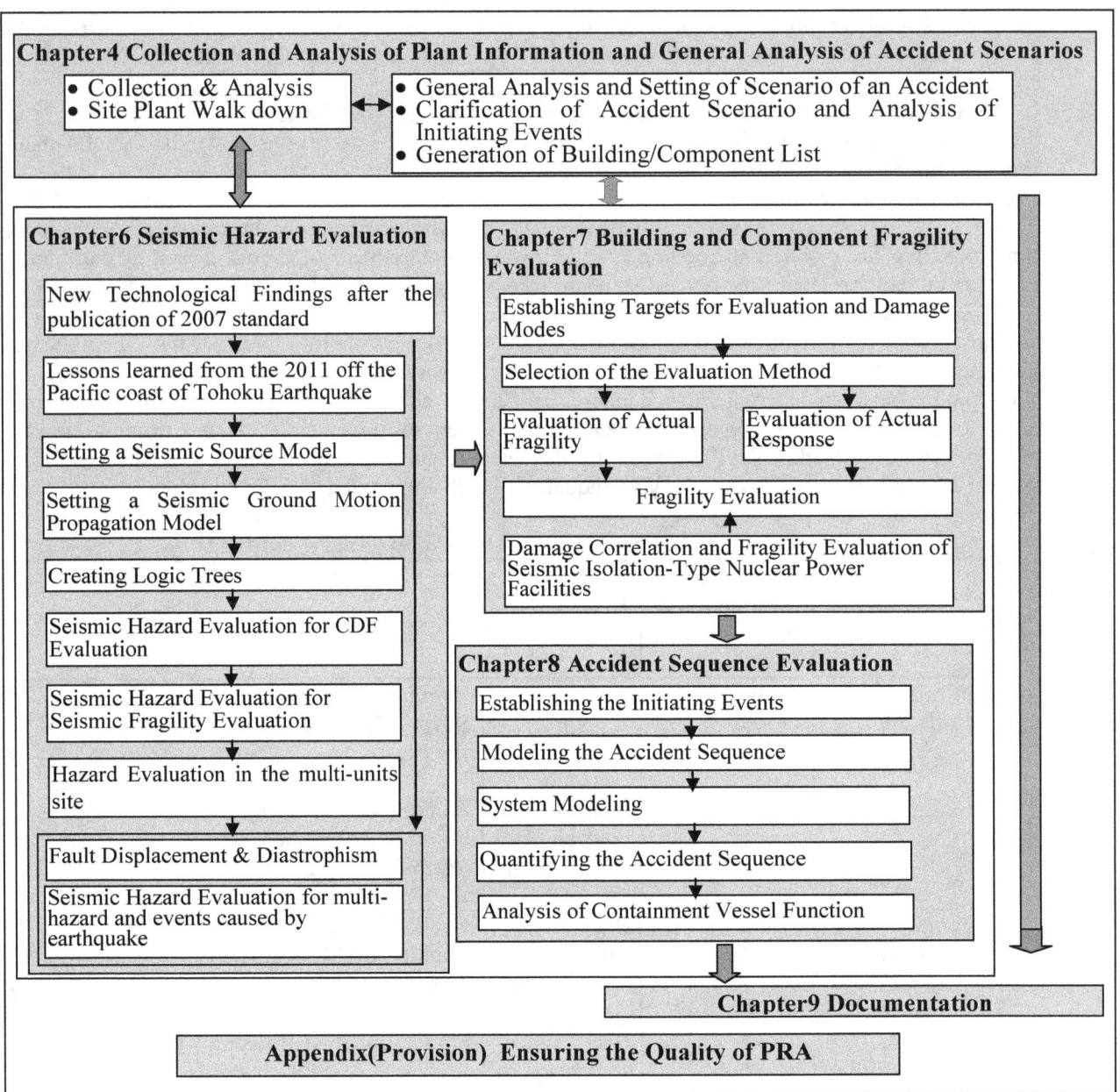

Figure 1 Evaluation Process (the revised Seismic PRA standard)

6. Collection and Analysis of Plant Information
6.1 Procedure of Collection and Analysis of Plant Information

Requirements related to the collection of plant information and the general analysis of accident scenarios are provided in Chapter 5. Outline of the process is Figure 2 as below.

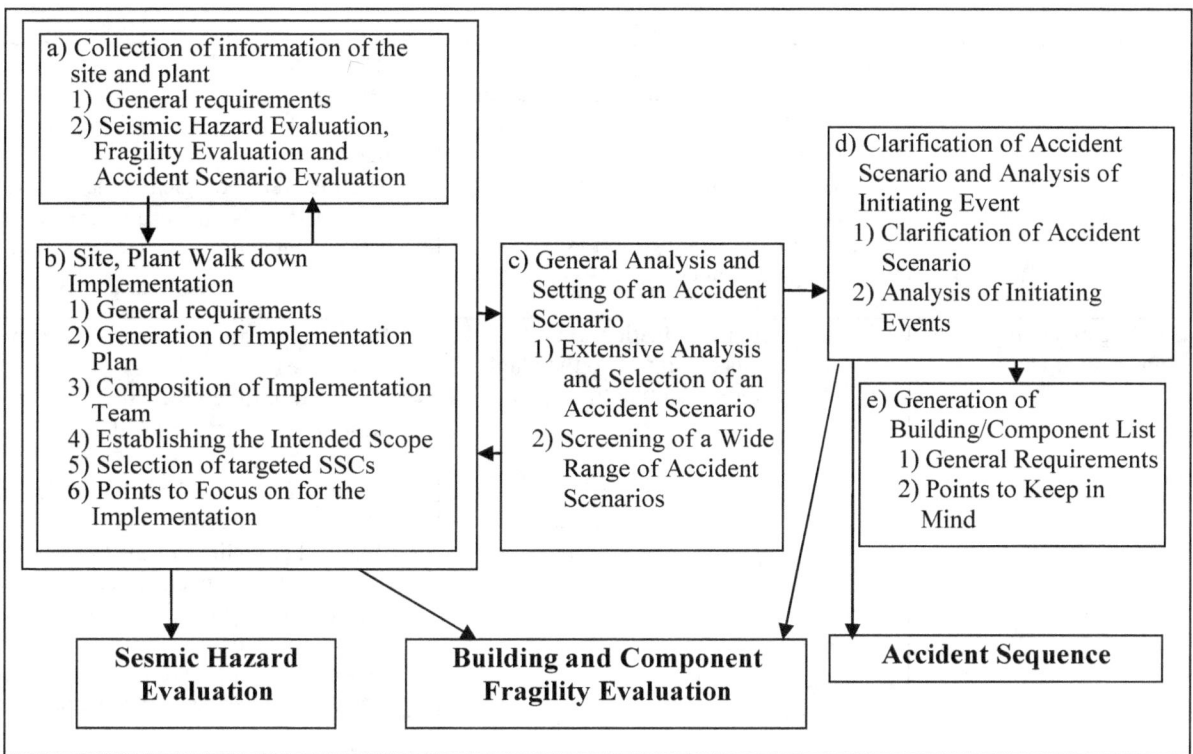

Figure 2 Flow of Collection of Information and Analysis of Accident Scenarios

6.2 Collection of information of the site and plant

Giving sufficient scrutiny to the scope of the information gathered and the amount of detail is provided in case of collecting and analyzing information of site and plants. Next, in accordance with the scope of the information collected, it is provided to check related information in such a way that recent plant conditions, operation experience and new knowledge. Especially, it is important to gather and analysis carefully the information and findings related to recent earthquake disaster including the 2011 earthquake off the Pacific coast of Tohoku. In addition to that, information and documents related to circumstances caused by combined with earthquake and tsunami, situation of components for accident management measures, mutual effects among plants in the same site, and effects of aftershocks.

In addition to information related to unique plant designs, operation, management and operation experience, collect a wide range of general information such as information related to existing seismic PRAs. Analyze the collected information to see whether it is sufficient from the perspective of reliability; if insufficiencies are found, collect additional information. When collecting additional information, implement additional inspections or tests as necessary.

When information that is not unique to plants is used in evaluation, it must be shown to be appropriate and rational by analysis of similarities and differentiae between general information and unique one. Essential information for seismic PRA is listed in Table 3 below.

Table 3 Essential information for seismic PRA

Evaluation Work	Required Information	Major Information
Understanding the design and operation of the plant	and operation management needed to perform the PRA • basic specifications • structural characteristics of the system facilities • characteristics of the seismic design • characteristics of the plant layout	• Basic plant specifications • Configurations and characteristics of system components • Seismic design features • Plant layout features • Various operating procedures • Domestic and foreign examples of seismic damage
Seismic hazard evaluation	Consider earthquake occurrence mode in the area surrounding the target site.	• Nuclear reactor facility permit applications

Evaluation Work	Required Information	Major Information
	Information on • seismic source characteristics that can be used to establish the seismic source model • seismic motion propagation characteristics that can be used to establish the seismic motion propagation mode	• Active fault and historic earthquake catalogs • Handbook of Japanese seismic fault parameters • Seismic area geological structure map • Records of seismic observations for the assessment site
Building and component fragility evaluation	Ultimate strength evaluation of buildings and components belonging to the plant, and information pertaining to response evaluations	• Permit application for installation of a nuclear reactor • Application for construction permit • Design /construction standards • Technical Guidelines for Seismic Design of Nuclear Power Plants by the Japan Electric Association • Relevant guidelines of the Society of Mechanical Engineers, Architectural Institute of Japan and the Society of Civil Engineers • Seismic design data, past test results, examples of earthquake damage, etc.
Accidents sequence evaluation a) Analysis of accident scenarios and classification of the initiating events	Plant conditions hypothesized at the time of a large scale earthquake	
b) Analysis of accident sequences • Establishment of the success criteria • Creation of event trees	• Conditions for use of systems such as the safety system • Realistic performance of systems • Mitigating operations undertaken by operators • Component failure mode and operation status for the target plant • Information that can be used to confirm the validity of evaluation results	• Realistic performance evaluation reports for systems related to the success criteria • Inspection procedures following an earthquake • Operating procedures (operating procedures for each facility, operating procedures for use during accidents, severance procedures) • Periodic inspection instructions • Training program for operators • Past PRA reports and other reports related to these following an earthquake
c) System modeling		
d) Quantification of the accident sequences		
e) Evaluation of containment integrity	Information on the isolation functions of the containment vessel	

6.3 Site-Plant Walk-down Implementation

The purpose of the walk-down for seismic PRA is collection of information that is hard to gather by paperwork. It is fine to carry out walk-downs multiple times as needed, but clarify the interdependent positioning of each (component) and the connections during each walk-down, for the walk-down's effective execution.

1) Generation of Implementation Plan
It is necessary to make implementation plan including list of walk-down team members, scope, targeted SSCs, procedures, and points of walk-down. In accordance with the purpose of each walk-down, it is possible to arrange the plan.

2) Composition of Implementation Team
The specialized ability, knowledge and experience are needed to the members of walk-down team. However anyone of members doesn't need to have all ability, knowledge and experience. Contents are as follows:
- Related to the systems, safety design and earthquake-resistant design for the plant targeted for assessment
- Related to vibration tests and seismic damage investigations related to the behavior of the components when there is seismic ground motion, as well as the damage sites and modes
- Related to seismic hazard evaluations, fragility evaluations (buildings, structures and components) and accident sequence evaluations

3) Establishing the Intended Scope
Items to keep in mind related to the establishment of the intended scope are indicated below.
- Include SSCs in the intended scope, where it is judged that they can not be assessed with the (already) collected information.
- It's useful to use past seismic PSA results to add SSCs.
- It's fine to exclude from the walk-down components for which fragility is clearly minor when compared to other components.
- If grasping the overall level of risk to the plant, focus on safety-critical components.
- If verification of the fragility of individual components that would have an impact on overall plant risk, it's focus on components for which a determination is made that the impact will be significant.
- Include SSCs that are common among plants or adaptable in case of accident management.
- Include SSCs that are evaluated in complicated PRA, e.g. a fire caused by earthquake PRA.

4) Points to Focus on for the Implementation
From the standpoint of fruitfulness, carry out a site-plant walk-down, focusing on the following.
- Safety Verification of Earthquake-Resistance - If it is judged that further information needs to be added within the design information necessary for the fragility assessment, carry out a review and verify the structures and components subject to assessment.
  - Compare the design documentation (system layout diagrams, instrumentation, piping system diagrams, single-line wiring connection diagrams, etc.) to the actual state of the plant, verifying the points that were judged to be insufficient in the information collected from paperwork. In particular, carry out a focused investigation and verification of the foundation sections of the subject components.
  - If there are items with relatively less probability of damage than other structures and components, which you can't decide whether to include in the assessment, verify the fragility of those items.
- Mutual Interference Between Components Due to Seismic Ground Motion – Verify unique characteristics of the plant such as mutual interference between components, mutual interference between systems and dependencies between systems.
- Verification of Secondary Impacts - Focus on verifying the secondary impacts of interference and collisions triggered by deformation, dislocation and movement through damage to components that are in functionally subordinate relationships.
- Verification of Accessibility After an Earthquake - Upon system assessment, verify accessibility when obtaining credit for components that need to be started-up on site and for components that can be expected to have functions restored with the recovery work on-site,.

7. General Analysis and Setting of an Accident Scenario

7.1 General Analysis

It's provided to analyze and set the scenario of an accident, using plant related information and information obtained in the site-plant walk-down. In analyzing and selecting a broad range of accidents, distill down specific accident scenarios at the time of an earthquake and select them, without overlooking any. For that purpose, it's necessary to consider the main factors such as follows:

- It's necessary to consider huge seismic ground motion occur several kind to components at the same time including not only preventive systems but also mitigation systems.
- It's meaningful to consider accidents that are directly linked to core damage by failure of reactor vessel or reactor building.
- It's necessary to analyze and select secondary accident scenarios that are not directly linked to the reactor core's damage, and where the damage exerts a direct impact on the damage to SSCs critical to safety and are possibly indirectly related to the reactor core's damage.
- Furthermore, in this section, requirements based on new lessons such as seismic caused complex events, SFP, impact of aftershocks, fault Displacement.

7.2 Clarification of an Accident Scenario

The focus is on the following 3 items to be ordered and clarified the accident scenario where it has been determined that accident sequence assessment is required .

- Clarification of the events for which there is a significant probability of their occurrence at the time of an earthquake and which lead to nuclear reactor core damage (hereinafter named "Accident scenarios that are characteristic during earthquakes.")
- Adjustment of accident scenarios those are characteristic during earthquakes and accident scenarios being considered for the internal event PRA.
- Setting of the minimum range for the subject earthquake strength in the seismic PRA.

It is point to keep in mind to clarify the accident scenarios generated by the following three damages.

- Damage by seismic ground motion to large static components for buildings, structures, piping etc. that are important for safety
- Damage by seismic ground motion to components, etc., that exert critical, wide-ranging impacts on safety functions
- Secondary impacts that have an impact on safety functions

7.3 Analysis of Initiating Event

In the analysis of the initiating events, characteristics that are specific to earthquakes are considered. Initiating events are classified according to the following six aspects:

- Multiple initiating events are classified as one initiating event when the same kind of mitigating equipment is required, the progressions of the events are similar, and the impacts are similar even if those initiating events are distinguish one from other.
- When it is difficult to rigorously analysis accident scenarios, it is fine to simplify these conservatively.
- When the possibility of damage to SSCs is extremely small, and it is determined that the probability of the occurrence of the initiating event is negligible, it is fine to exclude these as initiating events.
- When there are multiple causes of the initiating event, when the contribution to the occurrence probability of the initiating event is extremely small compared to other causes and when it is determined that the initiating event is negligible, it is fine to exclude these as initiating events.
- If the dependency of the initiating events were considered in the previous internal event PRA, it is necessary to keep them in mind with the seismic PRA as well.
- In events involving damage to containment vessels directly from seismic ground motion, because often the initiating event leads to an early release of FP, they are clearly segmented from other events.

7.4 Generation of Building/Component List

The work of generation of building/component list consists of collection/analysis of the plant information and general analysis of accident scenarios, the building/component fragility evaluation, and the accident sequence evaluation.

In the collection/analysis of plant information and the general analysis of the accident scenarios, first, collect the plant walk-down information and plant related information. Next, along with analyzing/setting a wide range of accident scenarios, screening those scenarios are implemented based on this information. Targeting the accident scenarios that remain from the screening, analysis of the initiating event and clarification of the accident scenario are implemented. These results create a target building/component list for the seismic PRA.

In the building/component fragility evaluation, structural screening of the evaluation events is carried out based on a damage mode analysis and a categorization of piping/components in the establishment of the subject of the evaluation. Then this information in the adjustments is reflected to the building/component list.

In the accident sequence evaluation, the information of SSCs required for the modeling of the accident sequence in the ET or FT.

In generation of building/component list, it's necessary to keep in mind four items below.

- Selecting SSCs required to achieve the prevention of core damage sequence:
  In consideration of the items such as the features of the system configuration, it is fine to provisionally exclude a portion from the assessment as is shown next.
    - ➢ SSCs with strong fragility are excluded in a system that has a serial architecture, with multiple SSCs that are dependent/subordinate.
    - ➢ SSCs with weak fragility are excluded in a system that has a parallel architecture, with multiple SSCs that possess redundancy.
- Gaining an understanding of the relative importance between SSCs:
  In case where the relative importance between SSCs is understood by provisionally evaluating the area targeted for evaluation, based on representative component data or data used in a previous PRA representing the structural areas of SSCs with weak fragility, it's necessary to keep in mind that both the data that is applied from previous PSA s as a standard and the data of the representative components do not provide results on the non-conservative side. The note is clearly stated that in evaluation at a latter stage a needed modification can be carried out as a provisional assessment.

8. CONCLUSION

The revised Seismic PRA Standard is now (in March 2014) open to public inspection by the SC (Standard Committee). The standard has several remarkable points to provide upgraded seismic PRA method. First of all, this standard covers the all area on seismic PRA and includes not only seismic events but also events caused by earthquake. In case of implementation of a fire caused by earthquake PRA, seismic hazard evaluation method comes from this standard, accident sequence evaluation method is based on the fire PRA standard, and fragility evaluation method of SSCs (structure, system and components) is based on the fragility chapter of this standard. Next, important information and findings from Fukushima Dai-ichi accident are added to this revised standard.

Seismic PRA or events caused by earthquake PRA can provide a lots of important and useful risk information to improve safety level of a nuclear power plant. After this revision the RTC ahs a plan to improve this seismic standard to implement a seismic shutdown PRA and a seismic at-power level 2 PRA.

References

[1]   AESJ, *"A Standard for Procedures of Seismic Probabilistic Safety Assessment for nuclear power plants,"* AESJ-SC-P006, (2007).

Revision of the AESJ Standard for Seismic Probabilistic Risk Assessment (2) Seismic Hazard Evaluation

Katsumi Ebisawa[a], Katsuhiro Kamae[b], Tadashi Annaka[c], Hideaki Tsutsumi[d] And Atsushi Onouchi[e]

[a]Tokyo City University, Tokyo, Japan
[b]Kyoto University, Kyoto, Japan
[c]Tokyo Electric Power Services Co., Ltd., Tokyo, Japan
[d]Former Japan Nuclear Energy Safety Organization, Tokyo, Japan
[e]Chubu Electric Power Co., Inc., Nagoya, Japan

Abstract: After the Atomic Energy Society of Japan was established seismic PRA implementation standard in 2007, some severe earthquakes which affect the seismic design of nuclear power plant have occurred. The most important earthquakes among them are the 2007 Niigata-ken Chuetsu-oki earthquake and the 2011 Tohoku-oki earthquake. In the later, the various new findings about the trigger earthquake and large aftershock caused by huge earthquake, the fault displacement and diastrophism due to the co-seismic and post-seismic slip, the joint effect of seismic motion and tsunami, and the effects of multi units and sites on the safety analysis were obtained. The new findings are incorporated into the revision of seismic hazard evaluation. This paper describes the overview of the Fukushima Dai-ichi nuclear power plant accident and lessons learned from its accident. The paper highlights the additional items based on lessons learned from various earthquakes such as Tohoku and NCO EQs after the 2007 version standard.

Keywords: Seismic PRA, Seismic hazard, Huge earthquake, Large aftershock, Combination of earthquake and tsunami

1. Introduction

The Atomic Energy Society of Japan (AESJ) had already established and published the implementation standard for Procedure of Seismic Probabilistic Risk Assessment (PRA) for nuclear power plants (NPPs) on 2007 (the 2007 version standard) through the discussions at the Seismic PRA Subcommittee under the Risk Technical Committee of the Standards Committee [1]. We had lessons learned from some earthquakes after the 2007 version standard. In particular, the lessons learned and new findings from the severe accidents of Fukushima Dai-ichi NPP (F1-NPP), which caused by Great East Japan Earthquake (Tohoku EQ) occurred on March 11 of 2011, were significant. In addition, those of Niigata-ken Chuetsu-oki Earthquake (NCO EQ) on July 17 of 2007 near Kashiwazaki-Kariwa NPP (KK-NPP) were also significant.

The objective of this paper, Part 2 seismic hazard, is to evaluate the seismic hazard for an accident sequence evaluation based on the following paper Part 4. The seismic hazard of the 2007 version standard was defined as the relationship between seismic motion and its exceedance frequency. The evaluation procedure was composed of the following seven sections, i.e. (1) process of seismic hazard evaluation, (2) handling of vertical motion and uncertainty factor, (3) setting of seismic source model, (4) setting of seismic motion propagation model, (5) formation of logic tree, (6) seismic hazard evaluation and (7) formulation of seismic motion for building and component fragility evaluation.

In revising the 2007 version standard, the definition of seismic hazard is also added the relationship between fault displacement and its exceedance frequency. Then the evaluation procedure is revised based on the new technological findings such as fault displacement, diastrophism, combination of seismic and tsunami events, multi units and is extended to the ten sections.

This paper describes the overview of the F1-NPP accident and lessons learned from its accident. The paper highlights the additional items based on lessons learned from various earthquakes such as Tohoku and NCO EQs after the 2007 version standard.

2. Overview of Fukushima NPP accident and lessons learned from Fukushima accident [2], [3]

2.1 Overview of F1-NPP accident

The F1-NPP is a multi-unit site with 6 BWRs as shown in **Fig. 1 (a)-(c)**. Figure 1(a) shows the location of each unit. Figure (b) and (c) are the cross and plan sections of reactor building (R/B) and turbine building (T/B) respectively. T/B stands directly by the sea. The emergency diesel generators are installed in the basement of these turbine buildings.

F1-NPP was overwhelmed by a tsunami about 46 minutes after the earthquake as shown in **Fig. 2**. The arrival time and tsunami height of the first large wave was 41 min after the main shock and O.P. of about 4 m, respectively. The arrival time and tsunami height of the second large wave were 8 min after the first wave with wave height. The tsunami height was so high that the experts estimated it to be more than 10 m from a photograph showing the overflow status of tsunami seawall (10 m) in **Fig.2**.

As to the sea water pump facilities for component cooling, all units were flooded by the tsunami as shown in **Fig. 2**. The Emergency Diesel Generators and switchboards installed in the basement floor of the reactor and the turbine buildings were flooded except for Unit 6, and the emergency power source supply was lost. Failure of reactor core cooling resulted in core damage in about 5 or 6 hours. Temperature and pressure in the primary containment vessel rose up, and radioactive materials were released through seals into the power plant and then the surrounding area. Consequently, a wide area was contaminated by the radioactive materials.

2.2 Lessons learned from the F1-NPP accident

The important issues of seismic engineering based on lessons learned from F1-NPP accident and Tohoku EQ are as follows [3];

(a) Occurrence of huge main earthquake and tsunami, a combination of seismic hazard and tsunami hazard,

(b) Consideration of huge aftershock and triggered earthquake,

(c) External events risk evaluation at multi units and sites,

(d) Combined emergency of both natural disaster and the nuclear accident,

(e) Core damage over a short period of time based on functional failure of support systems (seawater supply, power supply and signal systems),

(f) Common cause failure of multi structures and components and

(g) Dependency among neighbouring units.

The contents related to the issues from (a) to (c) will be found in chapter **3 to 5** later.

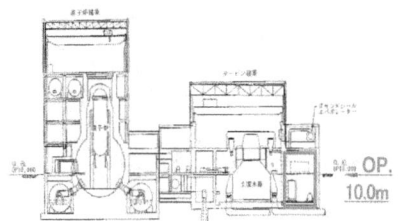

(b) Cross section of Reactor Building (R/B) and Turbine Building (T/B))

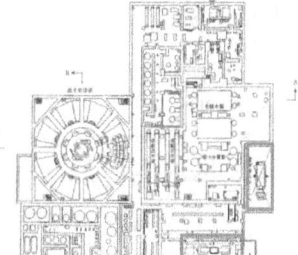

(c) Plan section of R/B and T/B

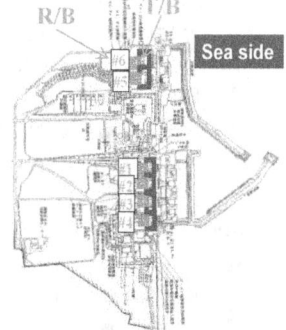

(a) Location of F1-NPP

Fig. 1 (a)-(c) Location of Fukushima Dai-ichi nuclear power plant

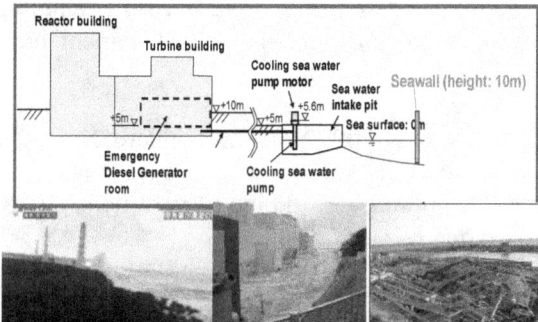

Fig. 2 Illustration of sea water supply system and situation of tsunami disaster at Fukushima Dai-ichi nuclear power plant (by Tokyo Elec. Power Co., 2011)

Probabilistic Safety Assessment and Management PSAM 12, June 2014, Honolulu, Hawaii

3. Policy for revising seismic hazard technology and additional items based on policy

3.1 Policy for revising seismic hazard technologies
The policy for revising seismic hazard technology is as follows;
(1) To analysis lessons learned from domestic and overseas some earthquakes after the 2007 version standard, to identify the important issues, and to consider them for the revised seismic hazard evaluation technologies,
(2) To analysis in detail especially lessons learned from NCO EQ (2007, Japan), Tohoku EQ (2011, Japan) and Aquila EQ (2010, Italy),
(3) To consider the consistency regarding the characteristics between seismic and tsunami sources.
(4) To consider the requirement from nuclear regulatory body based on F1-NPP accident and
(5) To describe in detail the examples that the 2007 version standard was applied to the safety inspection against NPP.

3.2 Additional items based on policy [3], [4]
The main additional items based on the above policy are as follows.
(1) The main target earthquakes are as follows.
- Domestic EQ: NCO EQ (2007), Iwate/Miyagi Prefecture EQ (2008), Tohoku EQ (2011) etc.
- Overseas EQ: Sichuan EQ (2008, China), Aquila EQ (2009, Italy), Christchurch EQ (2011, New Zealand) etc.
(2-1) Additional items From NCO EQ
- Treatment of stress concentrating zone   (6.3)
- Hazard considering multi units   (6.8)
(2-2) Additional items from Tohoku EQ
- Setting of source parameter of huge EQ   (6.3)
- Treatment of trigger EQ caused by huge EQ   (6.3)
- Hazard of large aftershock by huge earthquake   (6.6)
- Hazard of Fault displacement   (6.9)
- Hazard of diastrophism by huge EQ   (6.9)
- Hazard by considering combination of earthquake and tsunami events   (6.10)
(2-3) Additional item from Aquila EQ
- Administration responsibility of seismic expert   (6.5)
(3) Consistency of tsunami hazard evaluation
- Consistency between seismic and tsunami sources   (6.10)
(4) Requirements of nuclear regulatory body based on Tohoku EQ
- Evaluation of seismic motion generated by extremely near source   (6.4)
(5) Application example on inspection using the 2007 version standard
- Seismic hazard evaluation at KK-NPP   (6.5)

The number in ( ) corresponds the section numbers described in chapter 4 and 5 later.

4. Procedure of seismic hazard evaluation

The procedure of seismic hazard evaluation is described in chapter 6 of seismic PRA implementation standard. This procedure is composed of 10 sections considering the above additional items as shown in **Fig. 3**. These sections are divided into 3 parts, i.e. evaluation related to seismic hazard (including section 6.1), seismic motion hazard evaluation (including section 6.2 to 6.8) and fault displacement hazard evaluation (including section 6.9 to 6.10).
The technical contents of each section are as follows.
    Section 6.1: Lessons learned from earthquakes after 2007 version standard and their reflection to procedure of seismic hazard evaluation
    Section 6.2: Reflection of lessons learned from huge earthquake, treatment of uncertainty and validation and verification of seismic hazard evaluation
    Section 6.3: Setting of source model
    Section 6.4: Setting of seismic motion propagation model
    Section 6.5: Generation of logic tree
    Section 6.6: Evaluation of seismic motion hazard at bed rock

Section 6.7: Generation of time history wave due to fragility evaluation
Section 6.8: Notice items regarding seismic hazard evaluation at multi units
Section 6.9: Evaluation of hazard regarding fault displacement and diastrophism
Section 6.10: Multi hazard evaluation regarding external events

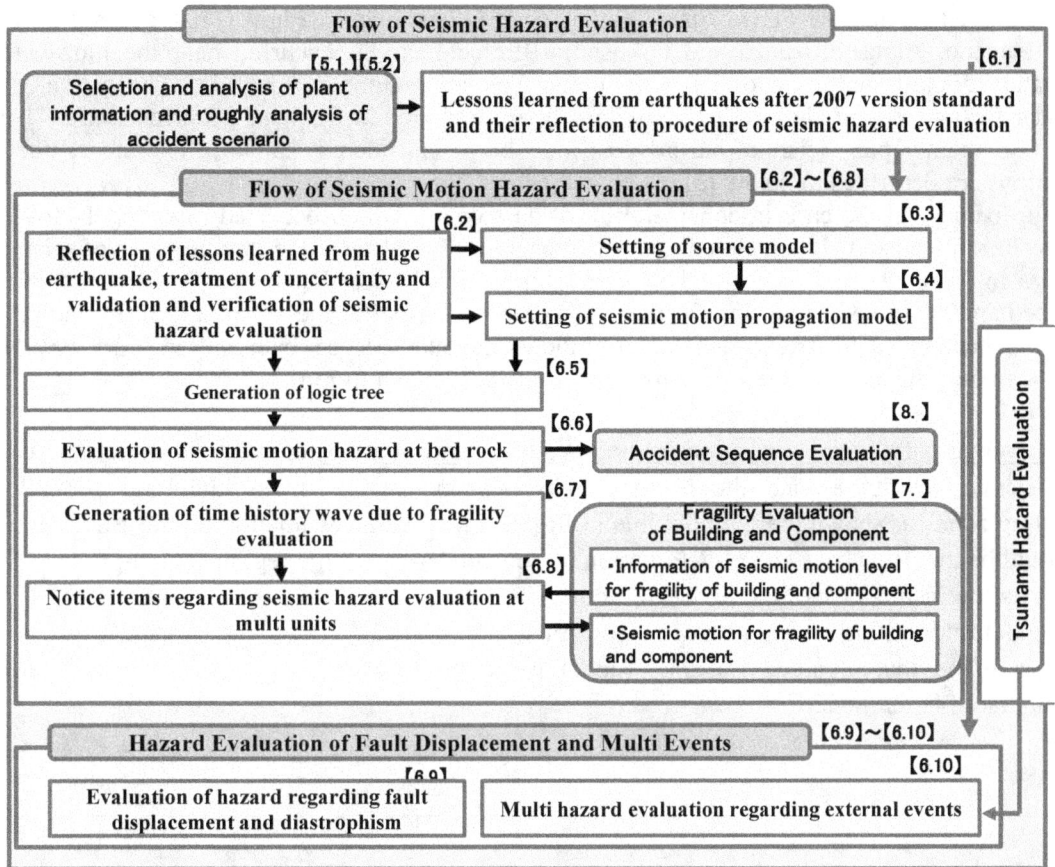

Fig. 3 Flow of seismic hazard evaluation

5. Additional items on each section

5.1 Additional items in section 6.1 "Lessons learned from earthquakes after the 2007 version standard and their reflection to procedure of seismic hazard evaluation

The additional items in section 6.1 are the following two them, i.e. (a) the contents of section 3.2 and (b) revised framework. In the above (a), the practical items are described in section 5.2 to 5.9 later. In (b), the practical framework is described as **Fig.3**.

5.2 Additional items in section 6.2 "Reflection of lessons learned from huge earthquake, treatment of uncertainty and validation and verification of seismic hazard evaluation"

The additional items in section 6.2 are the following two them, i.e. (a) Reflection of lessons learned from huge earthquake and (b) Validation & verification of seismic hazard evaluation. Here describes only (b).

In (b), it is described that the validation of seismic hazard evaluation is verified by referring the following evaluation example. This example compares the seismic motion level observed during time in a target area with the seismic motion level in seismic hazard curve corresponding to during time.

5.3 Additional items in section 6.3 "Setting of source model"

The additional items in section 6.3 are the following three them, i.e. (a) Setting of seismic source parameters for huge earthquake, (b) Treatment of triggered earthquake, (c) treatment of stress concentrating zone.

5.3.1 Setting of parameters for huge earthquake.

In the setting of seismic zone of huge earthquake, it is important to be not bound by preconceptions such as largest one in past data and to use imagination based on phenomena and physical investigation etc.

5.3.2 Treatment of triggered earthquake

The triggered earthquake (TE) caused by Tohoku EQ is shown in **Fig.4**. TEs occurred all over Japan including Nagano, Akita, Shizuoka and Fukushima Prefectures. TE occurred near the Idozawa fault belt approximately 50 km southwest of F1-NPP in the Tohoku region on April 11. The activated triggered earthquakes such as magnitude 6 to 7 approximately are included in frequent occurrence after Tohoku EQ. Influence of them upon seismic hazard has not been considered so far. Therefore, the expected considerations are described in below **[3]**.

It is probable that the case in consideration of TEs or not which occurred after the Tohoku EQ have different values of "a" and "b" on Gutenberug-Rihiter (G-R) Equation at the targeted area of seismic hazards evaluation. In order to confirm trend of probability, firstly, earthquake occurrence records can be accumulated and analysed in focusing for more ten years at least after the Tohoku EQ. Secondly, the values of "a" and "b" can be calculated by using data of the earthquake records for more ten years, besides trend of the values can be considered in view of before and after the Tohoku EQ **[3]**.

5.3.3 Improvement of b-value evaluation method in stress concentrating zone

The seismic activity around the NCO EQ hypocenter area is much high and so called "Stress concentrating zone" as shown the red bold line in **Fig. 5**. The b-value evaluation on the"Stress concentrating zone" should be modified based on G-R Equation on seismic hazard of the region source **[4]**.

Fig. 6 shows the results of b-value between modified b-value model and exiting b-value model. From this figure, it is found that b-value of former model is larger than that of latter model.

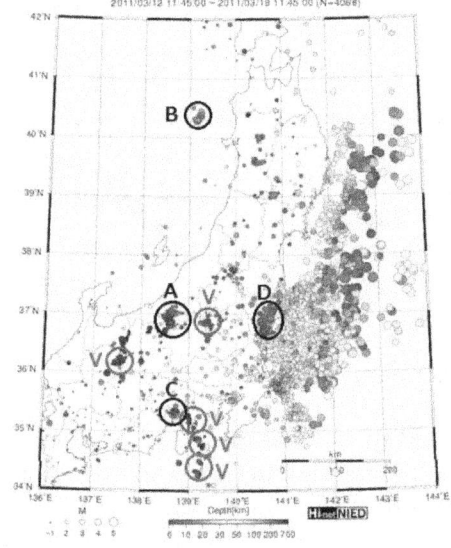

Fig. 4 Situation of occurrence of triggered earthquakes (symbol: ○) after 3.11 Tohoku earthquake

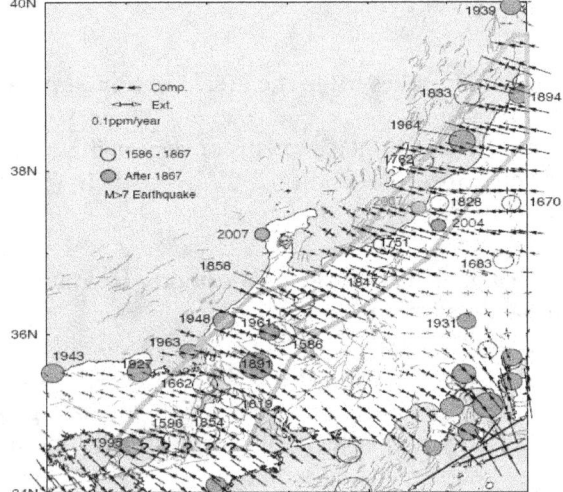

Fig.5 Example of stress concentrating zone around Niigata-ken Chuetsu-oki earthquake

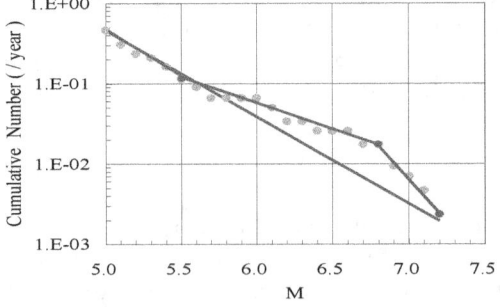

Fig.6 Example of the analysis results of b-value by both modified b-value model and exiting b-value model at stress concentrating zone

5.4 Additional item in section 6.4 "Setting of seismic motion propagation model"

The additional item in section 6.4 is the treatment for evaluating ground motions generated by extremely near sources. Nuclear regulatory body requires the above treatment.

Theoretical analyses show that the far-field terms are dominant in case of strong motion evaluations and the near- and intermediate-field terms are negligible. It follows that the simplification of the stochastic Green's function method, which neglects the so-called near- and intermediate-field terms, is valid for near-source strong motion evaluations [5].

The source model for an extremely near source, however, should reflect the complexity of the potential source rupture, especially the heterogeneous distribution of slip and rupture velocities. The rupture modelling method is good at characterizing these kinds of source effects and the seismic motions thus simulated are generally in a good agreement with the observation data even for those extremely near sources.

5.5 Additional items in section 6.5 "Generation of logic tree"

The additional item in section6.5 is the seismic expert responsibility related to the generation of logic tree. The background of this issue is as follows. In Aquila EQ (2009) in Italy, seismic experts were accounted for the administration responsibility.

It is described that the technical integrator, technical facilitator and experts take responsibility for only technical contents regarding seismic hazard evaluation. However they don't take one for the results of seismic hazard evaluation and safety of NPP based on the above contents.

5.6 Additional items in section 6.6 "Evaluation of seismic motion hazard at bed rock"

The cumulated number of aftershocks after Tohoku EQ is 6 for M greater than 7 as shown in **Fig.7**. M of the largest aftershock was 7.7 at 15:15 on March 11.

The additional item in section 6.6 is the treatment of huge aftershock hazard. The magnitude 9.0 of Tohoku EQ obeys relationship between fault length and magnitude as shown in **Fig.8**.

The concept of seismic hazard evaluation for huge aftershock is proposed as shown in **Fig.9**.

(1) For evaluation of seismic hazard for huge aftershock of M9 class EQs, confirmation should be made whether the main shock (M9) would follow the characteri~~~~ ~~~~ ~~~~~~~~ ~ ~ ~~~~~~~

(2) If it follows characteristics of the equation, obtain new C equation including the main shock (M9) as shown in Fig.

(3) Obtain occurrence frequency v (M9) of M9 using new C equation of (2).

(4) Obtain regression equation for aftershocks of M9 as sho in Fig. B.

(5) Obtain regression equation as conditional probability wit (M9) of (3) and regression equation of (4) as shown in F C.

(6) Obtain seismic hazard of aftershock using the regress equation of (5) as shown in Fig. D.

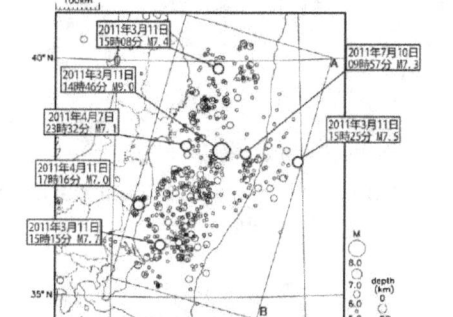

Fig. 7 Situation of occurrence of aftershock after 2011 Tohoku earthquake

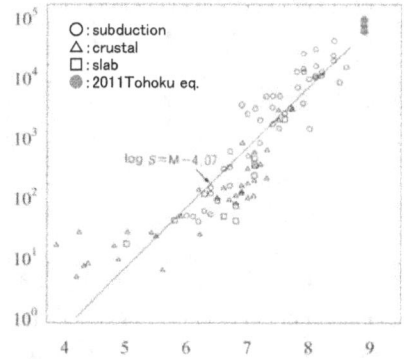

Fig.8 Relationship between fault length and magnitude including 2011 Tohoku earthquake

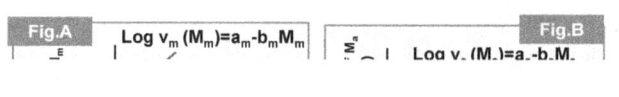

Fig.9 Procedure of seismic hazard evaluation for huge aftershock

5.7 Additional items in section 6.8 "Notice items regarding seismic hazard evaluation at multi units"

The additional item in section 6.8 is the seismic hazard at multi units and sites. NCO EQ occurred near KK-NPP. KK-NPP consists of 7 units as shown in **Fig.10**. In NCO EQ, the seismic motions that far exceeded those designed were observed at the building foundations of Unit KK1 to Unit KK7. In addition, the PGA at KK1 is about 2 times at KK5 because of the particular amplifying effect of irregular underground structure.

In seismic hazard evaluation, when seismic motions of all the target buildings and structures at a site are evaluated by using the same attenuation model, it shall be confirmed whether their seismic motions are similar value based on the seismic motion data observed at the site. If their data are not enough, its uncertainty factor needs to treat as the epistemic one. If their data are not different definitely, it is available to evaluate seismic motion by using the different attenuation model. In the fragility evaluation, it is advisable to confirm the response correlation between the target buildings and structures as shown in **Fig.11**.

5.8 Additional items in section 6.9 "Evaluation of hazard regarding fault displacement and diastrophism"

The additional items in section 6.9 are the following fault displacement and diastrophism hazards. In the former, a methodology for probabilistic fault displacement hazard evaluation was proposed by Youngs in 2003. This method established the evaluation formula on the basis of the surface earthquake faults that appeared when the normal faults moved. The evaluation formula based on the surface earthquake faults generated by reverse and strike faults in Japan were proposed. As a result of model case evaluations, the proposed evaluation formula gave a prospect for applicability in Japan **[6]**.

In the above method, the exceedance frequency of fault displacement hazard is evaluated as the sum of the frequencies of principal faulting and distributed faulting as shown in **Fig.12**. The example of evaluation result is shown in **Fig.13**.

Fig.10 Location of Kashiwazaki-Kariwa NPP with 7 units at Japan

Fig. 11 Concept of Evaluation of response correlation

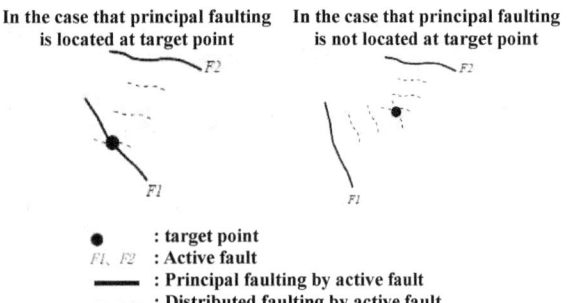

Fig. 12 Concept of displacement hazard evaluation

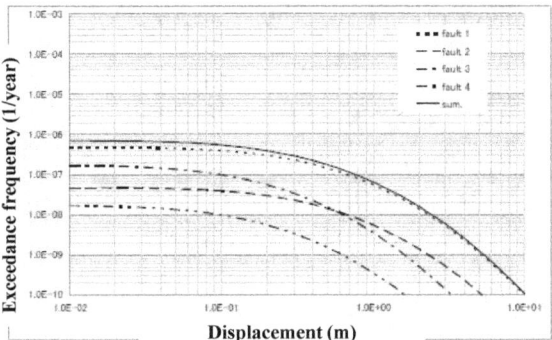

Fig. 13 Example of displacement hazard evaluation

5.9 Additional items in section 6.10"Multi hazard evaluation regarding external events"

The additional item in section 6.10 is the combination of seismic and tsunami hazards.

The seismic and tsunami hazard evaluations are practiced by developing hazard curves for seismic motion and tsunami height, respectively as shown in **Fig.14**. They are plotted against annual frequency of exceedance. Seismic hazard curves and tsunami hazard curves are not independent because they are based on common seismic events. But different nature of strong seismic motion (period range: 0.1~1sec) and tsunami rise time (period range: 10~120sec) requires careful consideration of their source characterization. Because of such difference in period ranges, correlated seismic motions at multi-unit locations should be considered, while tsunami height can be treated as more or less uniform within a single site **[3]**, **[7]**.

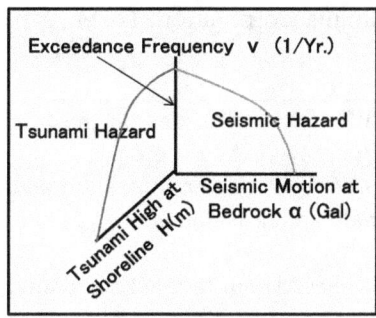

Fig. 14 Definition of hazard on seismic-tsunami PRA

6. Conclusion

This paper describes the overview of the F1-NPP accident and lessons learned from its accident. The paper highlights the additional items based on lessons learned from various earthquakes such as Tohoku and NCO EQs after the 2007 version standard.

Acknowledgements

This paper has drawn on significant contributions by participants in the seismic hazard evaluation working group and the seismic PRA subcommittee of the AESJ.

References

[1] Japan of Nuclear Atomic Society, "Seismic Probabilistic Safety Assessment Implementation Standards", (2009).

[2] Japanese government, "Report of Japanese government to the IAEA ministerial conference on nuclear safety", (2011).

[3] K. Ebisawa et al., "Current issues on PRA regarding seismic and tsunami events at multi units and sites based on lessons learned from Tohoku earthquake/tsunami", Korean Nuclear Society, Nuclear Engineering and Technology, Korea, VoL.44, NO.5, pp.439-452, (2012).

[4] K. Ebisawa, "Current status and important issues on seismic hazard evaluation methodology in Japan", Korean Nuclear Society, Nuclear Engineering and Technology, Korea, Vol.41, NO.10, pp.1223-1234, (2009).

[5] Japan Nuclear Energy Safety Organization, "Guidelines on seismic motion evaluation for design basis ground motion development: Methodology and its application to extremely near sources, JNES-RE-2013-2044, (2014).

[6] M.Takao et al., "Application of probabilistic fault displacement hazard analysis in Japan", Japan Association for Earthquake Engineering, Vol. 13, NO.1, pp.17-36, (2013).

[7] K. Ebisawa et al., "Concept for developing seismic-tsunami PRA methodology considering combination of seismic and tsunami events at multi-units", Proceeding of the International Symposium on Engineering lessons Learned from the 2011 Great East Japan Earthquake, Tokyo, Japan, pp.1575-1585, (2012).

Revision of the AESJ Standard for Seismic Probabilistic Risk Assessment
(3) Fragility Evaluation

Akira Yamaguchi[a], Susumu Nakamura[b], Yoshitaka Tsutsumi[c],
Tadashi Iijima[d] and Yoshinori Mihara[e]
[a]Osaka University, Osaka, Japan
[b]Nihon University, Koriyama, Japan
[c]Chubu Electric Power Co.,Inc., Nagoya, Japan
[e]Hitachi-GE Nuclear Energy, Ltd., Hitachi, Japan
[e]Kajima Corporation, Tokyo, Japan

Abstract: This paper introduces the following key issues on the fragility evaluation of SSCs in revision of the AESJ Standard for Seismic Probabilistic Risk Assessment.
1. Requirements for seismic induced other risk evaluations such as tsunami are clarified. For instance, the influence of structural damage due to main shock is considered as necessary to evaluate the realistic response by tsunamis after main shock.
2. Most recent findings are reflected based on the actual damage and simulation analyses of some earthquakes beyond design basis earthquake after 2007. For instance, seismic response analytical model is better suited for the realistic response evaluation up to damage limit paying attention to three dimensional responses of buildings / structures and its effect on equipment important to safety based on the seismic simulation analyses with observed records and usage experience. Floor deformation, torsion and rocking etc. are considered as three dimensional responses.
3. Requirements for the fragility evaluation of severe accident management equipment, its passageway, spent fuel pool and isolated important building are clarified based on the findings of Fukushima accident and so on.
4. Requirements for the fragility evaluation of aftershocks other than main shock and soil deformation due to fault displacement are clarified.

Keywords: PRA, Earthquake, Fragility, Standard.

1. INTRODUCTION

A standard for Procedure of Seismic Probabilistic Risk Assessment (PRA) for nuclear power plants 2007 had been already established and issued by the Atomic Energy Society of Japan (AESJ) through the discussions at the Seismic PRA Subcommittee under the Risk Technical Committee of the Standards Committee. As an enforcement standard based on the PRA procedure, the standard specifies the requirements which should have the PRA dealing with incidents resulting from earthquakes at nuclear power plants during power operation, and the concrete methods of meeting it.

In revising the 2007 version standard, we updated various requirements to reflect advancements in Seismic PRA techniques based on new technological findings after the publication of the previous standard and to improve the quality and transparency of this standard. In particular, the lessons learned and new findings from the severe accidents of Fukushima Dai-ichi nuclear power plants, which occurred on March 11 of 2011, were significant. The reason was that three cores were melted down and large amounts of FP were released in the accidents.

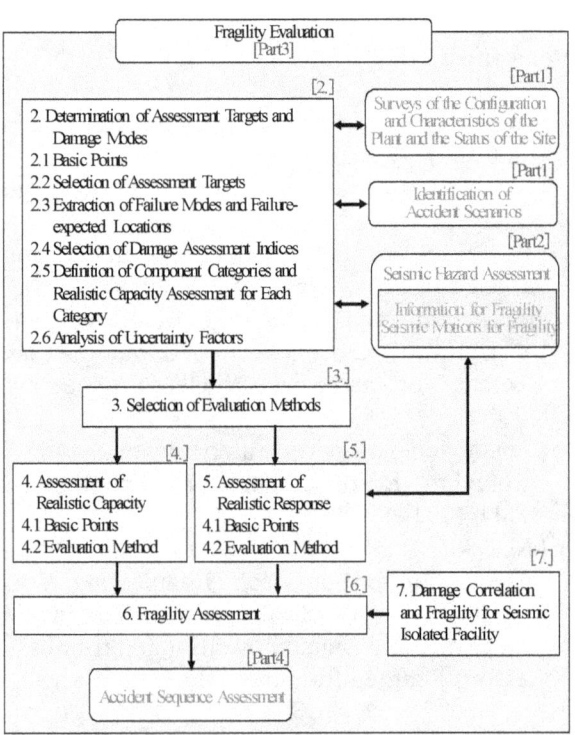

Fig.1 Procedures for Fragility Evaluation of Buildings and Components

The objective of this paper, Part3 fragility evaluation, is to evaluate the fragility of buildings and components for an accident sequence evaluation based on the following paper Prat4. For fragility evaluation of buildings and components, buildings and components to be assessed and the failure modes will be determined, and subsequently, the evaluation methods to be used for capacity evaluation and response evaluation will be selected to evaluate realistic capacity and response and thereby obtain fragility curves that show damage probabilities at which the response exceeds the capacity. This paper introduces the key issues on the above fragility evaluation of SSCs in revision of the AESJ Standard for Seismic PRA.

Fragility evaluation of buildings and components will be carried out according to the procedures shown in Fig. 1. In this revision, fragility curves not only for overall failure modes directly related to core damage but also for other local failure modes are strongly required if accident sequence evaluation needs the initiating events such as local SSC failures that consequentially influence core damage as well as the initiating events directly related to core damage such as reactor building collapse, reactor containment vessel collapse and reactor pressure vessel failure.

2. Determination of Assessment Targets and Failure Modes

2.1. Basic Points

The targets of fragility evaluation should be mainly selected on the basis of the lists of buildings and components extracted in "**Preparation of Lists of Buildings and Components**." Subsequently, dominant or potential failure modes and failure-expected locations should be extracted for the selected assessment targets. Damage assessment Indices should be also selected for fragility evaluation, appropriate to the selected failure modes and failure-expected locations.

Conditions for fragility evaluation specified here need to be arranged and shared with the evaluators of seismic hazard evaluation and accident sequence evaluation.

2.2. Selection of Assessment Targets

In addition to the above lists of buildings and components, the targets of fragility evaluation should also be selected on the basis of spent fuel damage outside reactor pressure vessel and other seismic induced PRA such as tsunami PRA. The targets except the lists include external and internal barriers against tsunamis, spent fuel pool and secondary equipment potential to become drifting articles by tsunamis after main shock.

2.3. Extraction of Failure modes and Failure-expected Locations

Failure modes and failure-expected locations to be assessed should be extracted by focusing on their failures that cause the direct and also indirect influence on the integrity of core, reactor containment vessel or spent fuel. Key issues in this revision are substantially as follows.

2.3.1. Buildings and Structures

The dominant modes of structural damage for the direct collapse (failure limit to support its own weight) and functional loss (equipment support functional loss and anti-leak functional loss etc.) of buildings and structures should be extracted. Not only failure modes such as overall collapse directly related to core damage but also other local failure modes should be evaluated in this revision. Fig.2 shows an example of the series of building failure modes from the viewpoint of combination between seismic and tsunami PRA.

2.3.2. Reactor Containment Vessel

Potential failure modes that are linked to the required functional loss in reactor containment vessel include overall structural collapse, structural failure modes due to functional loss of pressure resistance, functional failure modes due to loss of the containment vessel isolation and functional failure mode due to loss of the pressure suppressive function. From these, dominant failure modes and failure-expected locations to be assessed should be extracted.

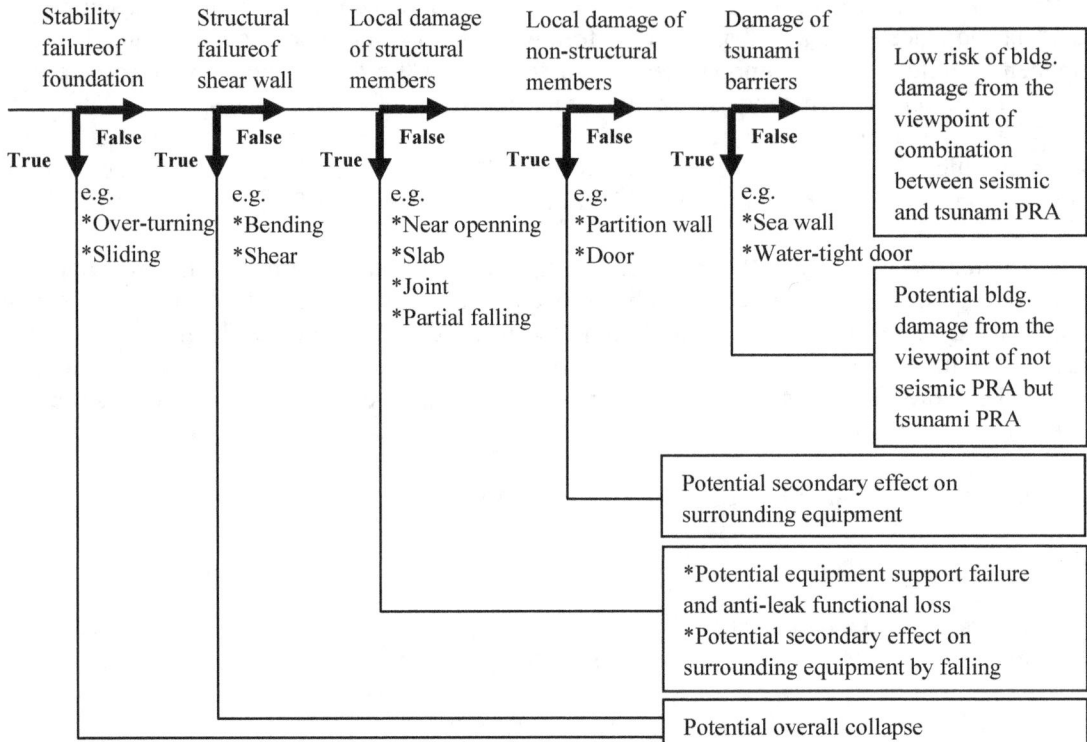

Fig.2 Example of the Series of Building Failure Modes

2.3.3. Components and Piping Systems

Failure modes and failure-expected locations to be assessed should be extracted based on the required functions of the item subject to evaluation. From the required functions of items subject to evaluation, failure modes are largely classified into two categories: structural failure modes and functional failure modes.

2.3.4. Soil

Soils to be assessed include foundation soil supporting the facilities important to safety, their surrounding slope and the passageway for severe accident management equipment. Failure modes and failure-expected locations to be assessed for each soil should be extracted based on the required functions of SSCs. Those for soil deformation due to fault displacement should be similarly extracted.

2.3.5. Tsunami Barriers

Potential structural failure modes that are linked to the required functional loss in external and internal barriers against tsunami such as sea wall include stability failure modes, overall structural collapse and failure modes due to their local damage. From these, dominant failure modes and failure-expected locations to be assessed should be extracted.

2.3.6. Spent Fuel Pool

Potential structural failure modes that are linked to the required functional loss in spent fuel pool include structural collapse and failure modes due to their local damage such as fracture of concrete and steel liner. From these, dominant failure modes and failure-expected locations to be assessed should be extracted. For example, bending or shear failure mode involving the appropriate story and the lower story collapse might be extracted as dominant failure modes and failure-expected locations that are linked to spent fuel damage.

2.3.7. Severe Accident Management Equipment

Failure modes are largely classified into two categories: structural failure modes and functional failure modes. Portable equipment such as power-supply car might be evaluated mainly for structural failure mode corresponding to over-turning against earthquake. That is because power-supply car is in storage during earthquake and will be used only for recovery work after earthquake.

2.4. Selection of Damage Assessment Indices

A realistic response quantity that can indicate the degree of damage in the target failure mode will be used as a damage assessment index. Damage assessment Indices will be selected appropriately from physical quantities used for describing the functional loss of buildings, structures and components due to seismic responses. As key issues in this revision, an example of damage assessment indices for various types of soils is shown in Table.1.

Table.1 Example of Damage Assessment Indices for Various Types of Soils

Soil		Damage assessment indices			Failure modes for SSCs
		Deemed limit states		Indecies necessary for fragility of SSCs	
		Instability limit for soil	Deformation limit for soil		
Foundation soil	Building	Safety factor for slip	Angle of slope of foundation	Movement distance of soil and rock mass	Structural or functional failure of buildings and switching station etc.
	Tsunami barrier	*Safety factor for slip/over-turning *Bearing capacity	-	-	Functional failure as tsunami barriers
Slope		*Safety factor for slip *Displacement of soil and rock mass	Displacement of soil and rock mass	Movement distance, volume and impact force of soil and rock mass	Structural or functional failure of buildings and switching station etc.
					Intake functional failure due to obstrction by falling soil to intake
					Functional failure as tsunami barriers of slope and dike
					reservoir functional failure by damage of reservoir bank
Soil for SAM equipment	Slope	*Bearing capacity *Safety factor for slip *Displacement of soil and rock mass	*Angle of slope of foundation *Displacement of soil and rock mass	Movement distance of soil and rock mass	Functional failure of SAM equipment
	Passageway	Safety factor for liquefaction	*Settlement and uneven distance *Displacement of soil and rock mass of slope	-	Functional failure of road surface as passageway
Soil deformation	Crustal movement	Safety factor for slip of slope	*Angle of slope of foundation *Displacement of soil and rock mass	Displacement of foundation soil	Structural or functional failure of buildings and switching station etc.
	Fault displacement			*Displacement of foundation soil *Movement distance of soil and rock mass	Structural or functional failure of building, switching station and underground structures etc.

2.5. Definition of Component Categories and Realistic Capacity Assessment for Each Category

In fragility evaluation of components, realistic capacity and realistic response are generally assessed for each component separately. However, categorization according to the structures, dimensions, shapes, operation mechanisms etc., of components sometimes makes it possible to carry out the same assessment and examination for all items in the same category. Accordingly, assessment and examination may be carried out for each category separately.

2.6. Analysis of Uncertainty Factors

For assessment of realistic capacity and realistic response, factors that have an influence on the probabilistic quantities (medians and standard deviations) of realistic capacity and realistic response (hereafter, collectively called "uncertainty factors") will be analyzed and extracted.

It is preferable that uncertainty be organized by dividing it into the following two categories to the extent possible: uncertainty due to randomness inherent in data or phenomena (aleatory uncertainty) and uncertainty related to knowledge and recognition in analytic techniques or modelling (epistemic uncertainty).

From the extracted uncertainty factors, major factors that have a significant influence on the finally obtained realistic capacity and realistic response may be extracted to assess realistic capacity and realistic response by using only those major factors.

3. Selection of Evaluation Methods

This standard basically presents the following three types of fragility evaluation methods.

    a) Method based on realistic capacity and realistic response [e.g.[1]]
    b) Method based on realistic capacity and response factor [2]
    c) Method based on capacity factor and response factor [e.g. [3]]

Fragility evaluation method for the realistic capacity and realistic response of SSCs should be selected depending on the application and accuracy required for the assessment. For selection of each evaluation method, any one of those techniques may be selected for use in the assessment, or a proper combination of several techniques may be used. A newly developed method except the above may be selected, but in such a case, the scientific rationality of the technique must be quantitatively shown.

4. Assessment of Realistic Capacity

4.1. Basic Points

Realistic capacity of SSCs should be evaluated for structural and functional failure modes of failure-expected locations. The following methods are presented in the standard.

    a) Method based on experiments
    b) Method based on empiricism including experiments
    c) Method based on theories including analyses
    d) Method based on engineering judgment

If only deterministic capacity is provided regardless of which method is used, realistic capacity with uncertainty should be evaluated by uncertainty analysis methods with material characteristics, etc. as aleatory variables.

4.2. Evaluation Method

This standard presents various evaluation methods and experimental data for realistic capacity of SSCs. As key issues in this revision, shaking table test data for realistic capacity of various types of components and piping systems are provided in the appendix. Furthermore, to evaluate realistic capacity such as soil deformation value, shaking table test data of scaled soil slope are also provided.

5. Assessment of Realistic Response

5.1. Basic Points

Realistic response of SSCs should be evaluated mainly based on the following two methods.

    a) Method based on realistic response
    b) Method based on response factor

Method based on realistic response is detail and exact one using new seismic response analyses. On the other hand, method based on response factor is relatively simple and approximate one using design response value.

As key issues in this revision, realistic response due to aftershocks other than main shock should be evaluated as necessary. Moreover, requirements for seismic induced other risk evaluations such as tsunami are clarified. For instance, the influence of structural damage due to main shock is considered as necessary to evaluate the realistic response by tsunamis after main shock.

5.2. Evaluation Method

This standard presents various evaluation methods for realistic response of SSCs. Some examples of key issues in this revision are the following.

a) Most recent findings are reflected based on the actual damage and simulation analyses of some earthquakes beyond design basis earthquake after 2007. For instance, the following requirement is clarified based on most recent findings. Seismic response analytical model is better suited for the realistic response evaluation up to damage limit paying attention to three dimensional responses of buildings / structures and its effect on equipment important to safety based on the seismic simulation analyses with observed records and usage experience. Floor deformation, torsion and rocking etc. are considered as three dimensional responses.

b) Not only indirect slope stability evaluation for effects on facilities due to seismic induced slope failure but also direct evaluation with sliding soil mass after failure and its impact force acting on facilities are specifically required.

c) Requirements for the fragility evaluation of soil deformation due to fault displacement are clarified.

6. Fragility Evaluation

Fragility curve should be calculated by the evaluation methods for realistic capacity and realistic response selected in "**3. Selection of Evaluation Methods**". The results of fragility evaluation of SSCs are used in accident sequence evaluation. In addition, many examples of fragility evaluation of SSCs are provided in the appendix. In this revision, fragility curves not only for overall failure modes directly related to core damage but also for other local failure modes are strongly required if accident sequence evaluation needs the initiating events such as local SSC failures that consequentially influence core damage as well as the initiating events directly related to core damage such as reactor building collapse, reactor containment vessel collapse and reactor pressure vessel failure.

7. Damage Correlation and Fragility Evaluation of Seismic Isolation Facilities

Under strong seismic ground motion, multiple components are damaged at the same time and it is assumed that so-called common cause failure will occur. Because of this, the interrelationships and correlations of damage between multiple components should be considered in the accident sequence evaluation.

Also, the vibration characteristics according to the seismic isolation types should be considered and evaluated in fragility evaluation of seismic isolation facilities. In this revision, requirements for the fragility evaluation of the isolated important building is clarified based on the findings of Fukushima accident and so on.

8. CONCLUSION

Revision of the AESJ Standard for the Seismic PRA standard will be established by AESJ in near future. It is expect to become the support of the decision making process in the wide field of the thing such as a safety design, operation management, safety regulation.

Acknowledgements

This paper has drawn on significant contributions by participants in the Building and Equipment Fragility WG and the Seismic PRA subcommittee of the AESJ.

References

[1] M.Miake et al., *"Study on Uncertainties in Fragility Evaluation of NPP Buildings"*, Summaries of Technical Papers of Annual Meeting Architectural Institute of Japan, 21552-21553, pp. 1103-11106, (2005).

[2] Japan Atomic Energy Research Institute, *"Report of Seismic PSA of BWR Model Plant"*, JAERI-Research 99-035,(1999).

[3] R.P. Kennedy and M.K. Ravindra, *"Seismic Risk Analysis Applied to Nuclear Power Plants"*, 8WCEE, Vol.7, pp.173-180, (1984).

Seismic Quantification Enhancements for getting CDF/LERF Distribution from the Point Estimates Results

Ovidiu Coman[a]

[a] International Atomic Energy Agency

Abstract: Technical requirements of the standard ASME/ANS RA Sa-2009 for capability category 2 imply appropriate consideration of uncertainty and combination of random failures with seismic failures. The paper presents how to develop the plant state mean fragility from the point estimate results that includes random failures. The plant state *CDF/LERF* components corresponding to each acceleration range are divided by the corresponding hazard frequency resulting discreet points of the mean plant state fragility. Furthermore using relationships presented in Ref. [2] β_U and β_R can be recovered and full plant state fragility parameters are obtained. Finally *CDF/LERF* distribution is developed.

Keywords: Seismic PSA, Random Failures, Fragility.

1. INTRODUCTION

High quality PSA that can be used as basis for various risk informed applications should comply with the technical requirements of the standard [1]. Seismic PSA should consider combination of random failures with seismic failures and also should properly consider variability associated to seismic hazard and seismic fragility functions. To address these high level requirements *CDF/LERF* distributions should be calculated at sequence level separate for random failures and seismic failures and after that combined to obtain in sequence *CDF/LERF* distribution and finally union of all significant sequences *CDF/LERF* define the plant state distribution of *CDF/LERF*. Special quantification tools are needed to perform such seismic quantification analysis.

Importance and sensitivity analyses are conducted to identify significant contributors and accident sequences. For practicality these analyses are performed for the point estimate level (mean *CDF/LERF*) and normal PRA software can be used (e.g. CAFTA, RISKSPECTRUM, etc.). The paper presents how plant state mean fragility can be obtained from the point estimates results. Also it presents an iterative process for approximating plant state mean fragility with a lognormal fragility that convolved with the mean hazard curve produce same mean *CDF* as the point estimate. The iterative process display the level of approximations introduce by this method and its impact to the final results. The equivalent lognormal plant state fragility allows recovering β_U and β_R by analytical solutions developed by the author [2]. Furthermore the full variability of the plant state fragility is obtained allowing to propagate both hazard and fragility variability in seismic *CDF* distribution.

2. EXTRACTION OF THE END STATE MEAN FRAGILITY

It is a common practice in S-PSA to split the hazard range of interest in several acceleration bins. Hazard frequency is calculated for each bin and point estimate results *CDF/LERFs* are obtained for each acceleration bin in a similar manner as for internal events PSA. Final mean *CDF/LERF* is obtained by simply summation of results obtained for each acceleration bin. In this analysis consideration of random failures is straightforward and does not require special quantification tools. This step of analysis is carried out to support importance, sensitivity and ranking analyses aimed to identify the main contributors and significant sequences. Observing S-PSA point estimated results the author identified a way of extracting points belonging to the mean plant state fragility (as many as number of acceleration bins are used in the analysis). These points belonging to the mean plant state fragility are obtained by dividing *CDF/LERF* for each acceleration bin to the corresponding hazard frequency.

Furthermore the author developed analytical relation to develop lognormal fragility equations crossing/bounding plant state fragility points. If only seismic failures are considered it is possible to find a single mean lognormal fragility crossing the points obtained from the plant state mean *CDF/LERFs* (as described above). If random failures are considered the plant state fragility is not lognormal distributed anymore but can be closely bounded by lognormal fragilities and an equivalent lognormal plant state fragility can be obtained. The equivalent lognormal plant state fragility has the property to produce the same mean CDF/LERF (by convolution with the mean seismic hazard) as compared to direct point estimates results. The approximation can be graphically displayed how close bounded the mean plant state fragility (close range of β_C and A_m values).

There are many advantages of getting the equivalent lognormal plant state fragility function:

    a. First allows extraction of the plant *HCLPF* in various cases:
        1) considering seismic failures, random failures and human errors
        2) considering only seismic failures (same as in SMA) and compare both margin estimates
    b. Second allows recovery of the full plant state fragility function by recovering β_R and β_U from the mean plant state fragility. If the mean plant fragility is lognormal than we have analytical solution to extract β_R and β_U and is possible to expand to the full fragility variability [2] and after that to convolve with the hazard curves for getting CDF/LERF distribution.

All these are showed in the following sections using an illustrative example.

3. ILLUSTRATIVE EXAMPLE

3.1. Description of the Illustrative Example

The basis for the illustrative example is presented in Tables A1 to A3 of the Annex A. Table A1 describes significant sequences corresponding to all acceleration bins (resulted after importance, sensitivity and ranking analysis are performed). Some of the sequences contain only seismic failures since other sequences include random failures and/or human errors associated to recovery actions. Table A2 presents fragility parameters for seismic basic events. The conditional probability of failure for each basic event corresponding to each seismic bin (S1 to S6) is calculated based on fragility parameters presented in Table A2. Table A3 presents random failures and human errors that appears in the sequences presented in Table A1. Finally Table A1 presents partial seismic *CDF* values corresponding to each acceleration bin and the sum of partial *CDF* values build the total seismic *CDF*. Table A1 allows calculation of contribution of the random failures as well as the contribution of the human errors to the final results. For this illustrative example we get:

CDF-without RF	1.09E-5	
CDF-without HERR	7.00E-05	
CDF including RF and HERR	1.68E-05	CDF without HERR and RF = 6.5E-5
HCLPF with RF and HERR	0.33g	HCLPF without HERR and RF = 0.27g

Can be observed that CDF without RF is 1.55 less than the CDF when RF is considered. Also CDF without HERR (no recovery actions are credited) is 4.16 higher than CDF when HERR is considered. A qualitative observation shows that in case of SMA ignoring both random failures and human errors associated to recovery actions lead to a conservative estimate of seismic margin. Also it will be un-conservative to credit operator recovery actions and not considering combination of seismic failures with random failures.

3.2. Recover Plant State Lognormal Equivalent Fragility from the Point Estimate Results

The process for recovering the plant state equivalent lognormal fragility is described using the illustrative example presented in Annex A. Table 1 summarize results of the illustrative example containing:

- definition of each acceleration bin (S1 to S6),
- acceleration value within each bin for which fragility point associated to that bin is defined,
- hazard frequency for each acceleration bin,
- partial seismic *CDF* for each acceleration bin and total seismic *CDF,*
- plant state fragility points obtained by dividing partial seismic *CDFs* by the hazard frequency,
- relative *CDF* contribution of each acceleration bin.

It should be noted that partial *CDF* values and plant state fragility points include the effect of seismic failures, random failures and human errors.

Table 1: Calculation of Plant State Fragility Points based on Partial *CDFs*

| Bin | Frag acc. Point | Acceleration Bin | | H(a1) | H(a2) | Hazard. Freq. | Partial CDF | Plant State Frag. | Contrib. % |
		a1	a2						
S1	1.5E-01	0.1	0.2	5.53E-01	5.83E-03	5.47E-01	1.4E-07	2.6E-07	0.85
S2	2.2E-01	0.2	0.3	5.83E-03	4.07E-04	5.43E-03	1.0E-06	1.9E-04	6.13
S3	3.6E-01	0.3	0.4	4.07E-04	6.15E-05	3.45E-04	7.9E-06	2.3E-02	46.94
S4	4.7E-01	0.4	0.6	6.15E-05	4.29E-06	5.72E-05	4.8E-06	8.4E-02	28.64
S5	7.5E-01	0.6	0.8	4.29E-06	6.49E-07	3.64E-06	1.9E-06	5.2E-01	11.16
S6	1.0E+00	0.8	1.2	6.49E-07	4.53E-08	6.04E-07	1.1E-06	1.0E+00	6.27

Total Seismic *CDF*   1.68E-5

The hazards curves and relative contribution of each acceleration bin to the total *CDF* are presented in Figure 1. Figure 2 presents the plant state fragility (before and after fine adjustment of the fragility acceleration points within each bin). The adjustment of fragility acceleration point is done in order to reduce deviation from lognormal plant state fragility introduced by the random failures including the conditions that the mean CDF is conserved. The folowing equation is used to obtain lognormal fragility parameters crossing the fragility acceleratoin points corresponding to each acceleration bin:

$$A_{m-i} = \frac{a}{e^{\left(\Phi^{-1}(f(a))\beta_C\right)}} \tag{1}$$

where: a – correpsonds to fragility acceleration point for each bin *S-i* and *F(a)* is the plant state fragility for *S-i* (see Table 1). Using eqation (1), β_C is iterated until A_{m-Si} values (coresponding to acceleration bins with important contribution to *CDF*) get close enough to the plant state fragility points shwon in Table 1.

The iterative process is illustrated in Table 2 and Figure 3. Solution converges to lognormal fragility parameters that closely bound the plant state fragility points. The equivalent plant state lognormal fragility parameters are those that gives the same point estimated seismic CDF. In other words convolution between the mean hazard curve with the equivalent plant state lognormal fragility produce same seismic CDF as the one obtained by point estimate shown in Table 1 or Table A1 for Annex A.

Table 2 Numerical results of iterations for getting Plant State Lognormal Mean Fragility

| A_m | CDF Contrib. % | Bc Iterations 1 to 4 | | | |
		0.25	0.40	0.33	0.341
A_{m-S1}	0.85	5.26E-01	8.07E-01	7.86E-01	8.31E-01
A_{m-S2}	6.13	5.35E-01	6.78E-01	7.11E-01	7.40E-01
A_{m-S3}	46.94	5.93E-01	6.34E-01	6.96E-01	7.12E-01
A_{m-S4}	28.64	6.63E-01	6.25E-01	7.41E-01	7.52E-01
A_{m-S5}	11.16	7.43E-01	5.65E-01	7.41E-01	7.41E-01
A_{m-S6}	6.27	6.63E-01	5.62E-01	5.81E-01	5.70E-01

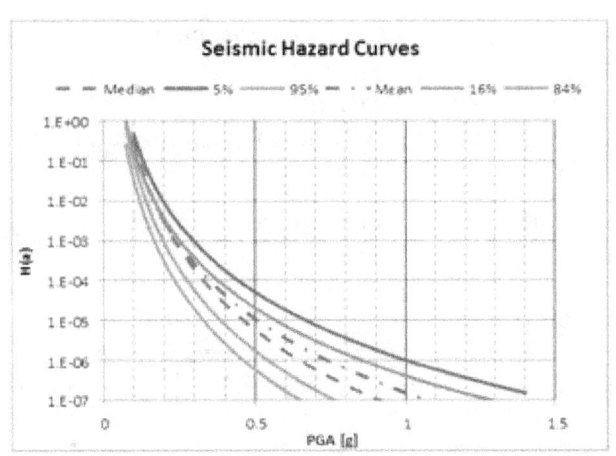
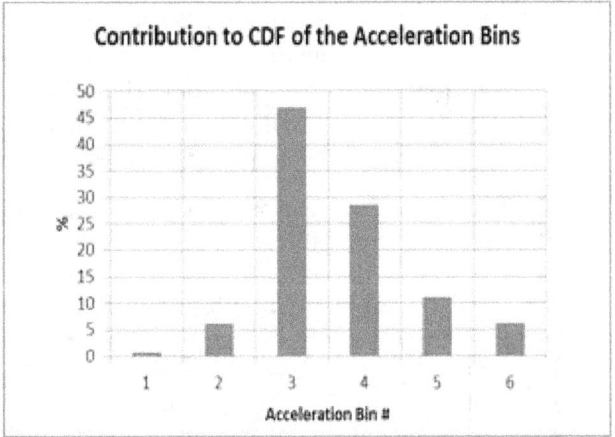

Figure 1 Seismic hazard curves and relative contribution of acc. bins to total seismic CDF

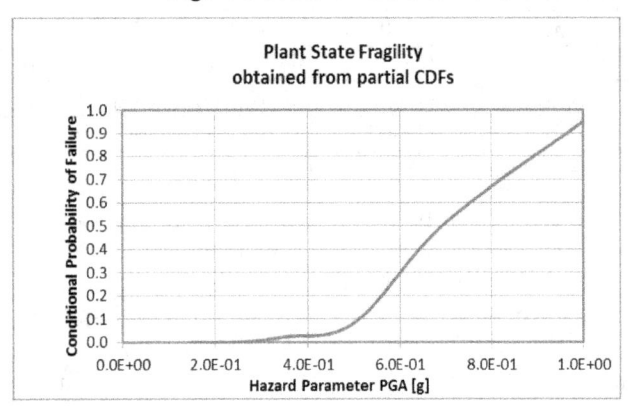
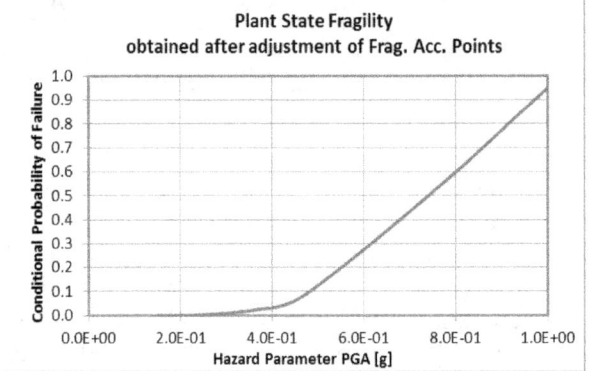

Figure 2 Plant state fragility obtained from partial CDFs divided by hazard frequency for each acc. bin.

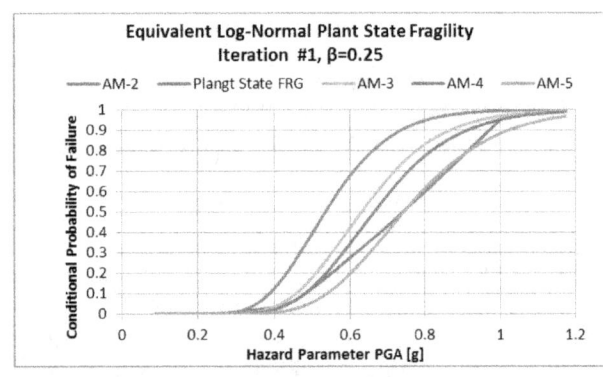
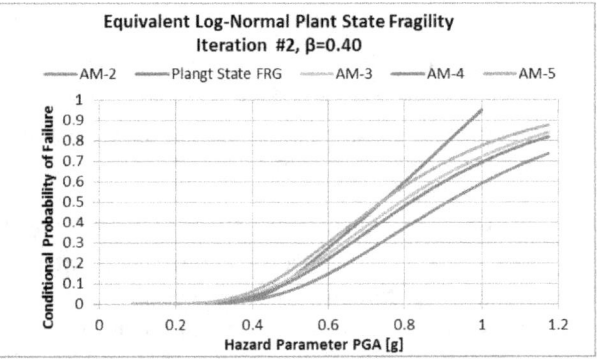
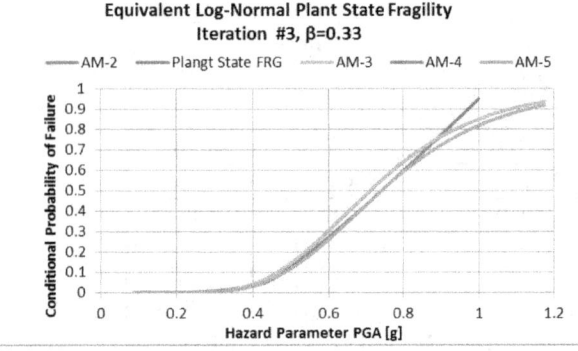
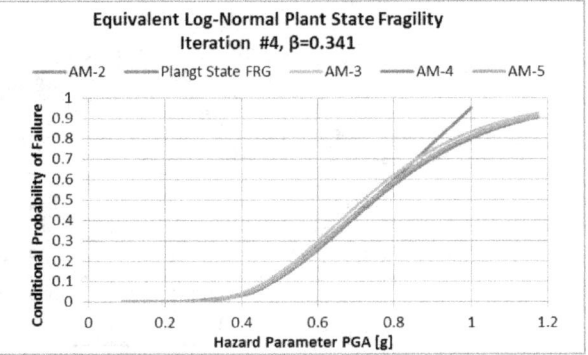

Figure 3 Graphic Results of Iterations for getting equivalent lognormal plant state fragility

The median capacity A_m corresponding to the equivalent lognormal plant state fragility is calculated as weighted sum of $A_{m\text{-}Si}$ for each acceleration bin corresponding to the last iteration. The relative acceleration bin contribution to CDF is used as weighting factors. In this example, using the proposed iterative process the following plant state lognormal fragility parameters have been obtained:

- A_m = 0.72g *and HCLPF*=0.325g, β_C =0.341 (including the effect of random failures).
- *CDF* = 1.68E-5 (by convoluting the derivative of the mean hazard and mean fragility) is the same as the one shown in Table 1 obtained by point estimate considering and random failures and human errors.

This shows that the equivalent lognormal mean plant state fragility obtained in this way is conserving the point estimate *CDF* value. Also Table 2 and Figure 3 shows numerically and graphically the approximation range (for A_m and β_C) introduced by this procedure.

3.3. Recover β_U and β_R from the Equivalent Plant State Mean Fragility

As presented in [2] analytical relationships have been develop to obtain β_R and β_U based on the mean fragility parameters (A_m, *HCLPF* and β_C):

$$\beta_U = \frac{B + \sqrt{B^2 - 4AC}}{2A} \qquad\qquad \beta_R = \frac{B - \sqrt{B^2 - 4AC}}{2A} \qquad (2)$$

$$A = 2 \qquad B = 2\frac{\ln\left(\dfrac{HCLPF}{A_m}\right)}{\phi^{-1}(0.05)} \qquad C = \left(\frac{\ln\left(\dfrac{HCLPF}{A_m}\right)}{\phi^{-1}(0.05)}\right)^2 - \beta_C^2 \qquad (3)$$

The equivalent lognormal plant state mean fragility parameters obtained in Sub-Section 3.2 are:

$$A_m = \quad 0.72 \qquad \beta_C = 0.341 \qquad HCLPF = 0.325$$

Using equations above plant state fragility parameters and equations (2) and (3) we get:

$$A=2 \qquad B=0.963 \qquad C=0.1152$$
$$\beta_U= \quad 0.258 \qquad \beta_R=0.223$$

Finally the equivalent lognormal plant state fragility function (considering all variability) is defined by the following parameters:

$$A_m = 0.72; \quad \beta_U=0.258; \beta_R= 0.223$$

The above described process lead to the equivalent lognormal plant state fragility function consistent to point estimate results that includes contribution of random failures and human errors. Plant state fragility function can be further used to calculate seismic CDF distribution.

3.4. Seismic *CDF* distribution

Figure 4 illustrates the required input needed for calculation of the seismic *CDF* distribution. Equation (4) can be used to develop a number of plant state fragility corresponding to different confidence levels "*Q*" (typical 25 to 50 fragility curves) – also associated probability distribution parameters are needed. Same number of seismic hard curves should be available from PSHA study or should be developed based on available PSHA results.

$$F(a) = \phi\left(\frac{\ln\left(\dfrac{a}{A_m}\right) + \beta_U \phi^{-1}(Q)}{\beta_R}\right) \qquad (4)$$

Each fragility curve is convolved with each hazard curve resulting the *CDF* value and corresponding distributions parameters. This pair of values defines one *CDF* distribution point. For all combinations of fragility and hazard curves the seismic *CDF* distribution is obtained as shown in Figure 4 and 5.

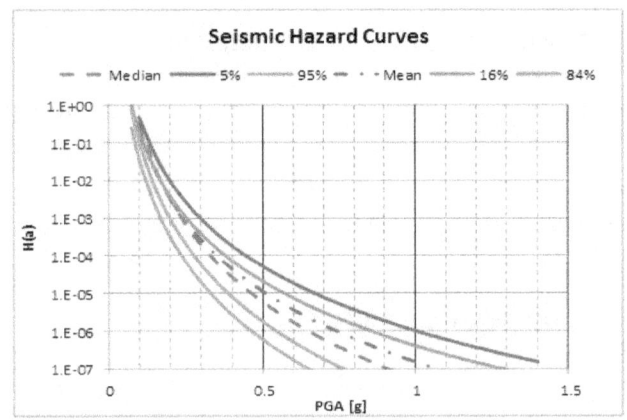 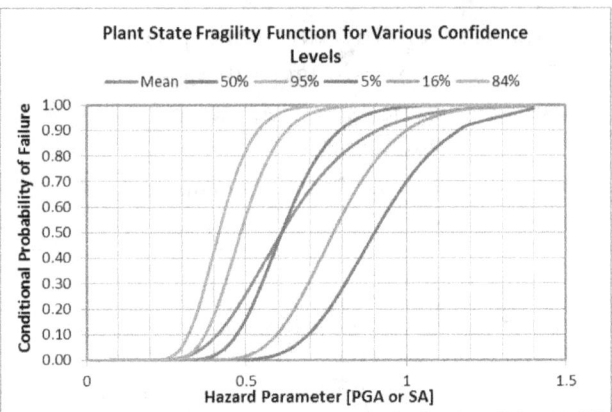

Figure 4 Input for Seismic Risk Quantification

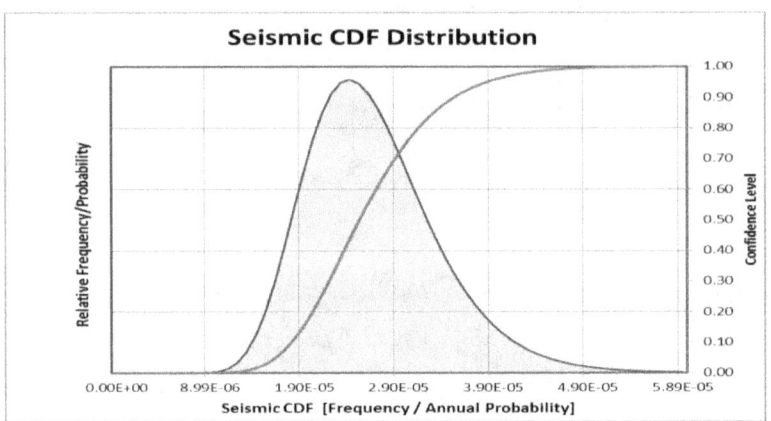

Figure 5 Seismic *CDF* Probability Distribution

4. CONCLUSIONS

The plant state point estimate of seismic *CDF/LERF* is obtained for each acceleration bin and summation of partial *CDFs/LERFs* give the total seismic *CDF/LERF* corresponding to the acceleration range of interest. Plant state fragility points are simply obtained by dividing partial *CDFs/LERFs* with the hazard frequency corresponding to each acceleration bin. Due to the random failures these points does not belong to a lognormal mean fragility.

Equivalent lognormal mean plant fragility can be defined for each acceleration bin using equation (1) and iterating β_C until these fragilities closely bound all plant state fragility points (that have significant contribution to CDF/LERF). The equivalent plant state lognormal mean fragility parameters are obtained as a weighted sum of median capacity values A_m for each acceleration bins and β_C values form the last iteration. Second condition used in the iteration process is that the equivalent lognormal end state mean fragility produce same mean *CDF* as compared with the one obtained initially for the point estimate analyses results presented in Annex A and Table 1.

The approximation introduce in this process can be displayed numerically (variation of A_m capacity for different acceleration bins) and graphically (bounding of plant state fragility points). The equivalent plant state lognormal fragility obtained at the end of this process has the property to conserve the mean seismic *CDF* and closely cross the plant state fragility points (corresponding to significant acceleration bins) that include the effect of random failures.

There are many advantages developing equivalent lognormal plant state mean fragility. One of advantage is that the lognormal fragility can be expanded to the full fragility function – recovering β_R and β_U using analytical relationships developed by the author [2] and finally the seismic CDF distribution can be obtained. Also seismic margin estimates can be easily developed including contribution of the random failures and human errors. The author believes that point estimates with these enhancements converge to results obtained by the accurate quantification process described at the beginning of Section 1.

5. REFERENCES

[1] External Hazard PRA Standard ASME/ANS RA-Sa-2009.
[2] O. Coman, "Some useful enhancements for Seismic PRA" ANS PSA 2013 International Topical Meeting on Probabilistic Safety Assessment and Analysis, Columbia, SC, Sept. 2013.
[3] IAEA/ISSC EBP-WA2 – Draft Technical Report on "Seismic Probabilistic Safety Assessment Implementation Guidelines" (Draft 3 January 2014).

Annex A

ILLUSTRATIUVE EXAMPLE

Table A1: Significant sequences/cutsets including random failures, human errors and partial CDFs for each acceleration Bin.

Seq. Freq.	Acc. Bin	H-Freq	S-IE	S-F1	SF-2	RF	HERR	DESC	Partial CDF
3.2E-09	S1	5.5E-01	1.0E-02	8.4E-04	3.4E-02		2.0E-02	A1*B1*HER1	CDF-S1
1.4E-07	S1	5.5E-01	1.0E-02	8.4E-04			3.0E-02	A1*B1*RF2	1.4E-07
4.6E-07	S2	5.4E-03	5.4E-02	4.6E-02	3.4E-02			A1*B1*E1	
1.5E-09	S2	5.4E-03	2.4E-02	2.6E-04			4.5E-02	T1*C1*HER2	
2.6E-07	S2	5.4E-03	2.4E-02	1.0E-02	1.9E-01			T1*A1*C1	CDF-S2
3.0E-07	S2	5.4E-03	5.4E-02	3.4E-02			3.0E-02	A1*E1*HER3	1.0E-06
7.0E-09	S2	5.4E-03	2.4E-02	1.9E-01	3.6E-03	7.5E-02		T1*C1*E3*RF3	
7.1E-07	S3	3.5E-04	1.6E-01	1.8E-01		7.3E-02		E1*A1*RF5	
2.2E-06	S3	3.5E-04	1.6E-01	5.4E-01			7.5E-02	E1*C1*HER5	
2.2E-06	S3	3.5E-04	1.9E-01	2.4E-01	1.4E-01			A1*B1*E1	
2.8E-06	S3	3.5E-04	2.0E-01	5.4E-01		7.5E-02		T1*C1*RF3	CDF-S3
1.7E-10	S3	3.5E-04	2.0E-01	1.6E-03	1.6E-03			T1*E2*C2	7.9E-06
9.4E-09	S3	3.5E-04	1.9E-01	1.6E-03	8.7E-02			A1*E2*B2	
1.7E-09	S3	3.5E-04	5.7E-03	1.6E-03	5.4E-01			M1*E2*C1	
1.7E-10	S3	3.5E-04	5.7E-03	1.6E-03			5.5E-02	M1*E2*HER4	
1.7E-06	S4	5.7E-05	3.4E-01	6.2E-01	1.4E-01			A1*B1*E1	
1.1E-06	S4	5.7E-05	4.3E-01	6.2E-01		7.5E-02		T1*C1*RF3	
9.6E-09	S4	5.7E-05	4.3E-01	2.0E-02	2.0E-02			T1*E2*C2	
1.4E-07	S4	5.7E-05	3.4E-01	2.0E-02	3.7E-01			A1*E2*B2	
3.9E-10	S4	5.7E-05	2.3E-02	2.0E-02		1.5E-02		M1*E2*RF4	
5.2E-10	S4	5.7E-05	2.3E-02	2.0E-02	2.0E-02			M1*E2*C2	CDF-S4
1.3E-09	S4	5.7E-05	2.3E-02	2.0E-02		5.0E-02		M1*E2*RF7	4.8E-06
1.1E-06	S4	5.7E-05	4.3E-01	6.2E-01		7.5E-02		T1*C1*RF3	
6.1E-08	S4	5.7E-05	2.3E-02	6.2E-01			7.5E-02	M1*C1*HER5	
6.5E-07	S4	5.7E-05	3.4E-01	6.2E-01			5.5E-02	E1*C1*HER4	
1.4E-06	S5	3.6E-06	6.4E-01	9.0E-01	6.8E-01			A1*B1*E1	CDF-S5
2.1E-07	S5	3.6E-06	8.4E-01	9.0E-01		7.5E-02		T1*C1*RF3	1.9E-06

Seq. Freq.	Acc. Bin	H-Freq	S-IE	S-F1	SF-2	RF	HERR	DESC	Partial CDF
3.8E-08	S5	3.6E-06	8.4E-01	1.1E-01	1.1E-01			T1*E2*C2	
1.9E-07	S5	3.6E-06	6.4E-01	1.1E-01	7.3E-01			A1*E2*B2	
1.8E-08	S5	3.6E-06	5.0E-02	1.1E-01	9.0E-01			M1*E2*C1	
4.3E-07	S6	6.0E-07	8.0E-01	9.9E-01	9.0E-01			A1*B1*E1	CDF-S6 1.1E-06
3.1E-07	S6	6.0E-07	9.6E-01	9.9E-01	5.4E-01			T1*C1*E3	
8.0E-08	S6	6.0E-07	9.6E-01	3.7E-01	3.7E-01			T1*E2*C2	
1.7E-07	S6	6.0E-07	8.0E-01	3.7E-01	9.5E-01			A1*E2*B2	
7.0E-08	S6	6.0E-07	3.1E-01	3.7E-01	9.9E-01			M1*E2*C1	
Total *CDF*									1.7E-05

Notes:

H-Freq = Seismic event frequency for acceleration bin #1, …6
S-IE = conditional probability of seismic initiating event
SF-1, SF-2 = seismic failures conditional probabilities
RF = Non-seismic random failure probability
HERR = Human error associated to operator's recovery actions

Table A2: Seismic Fragility Functions

BE	A_m	HCLPF	β_U	β_R	β_C
A1	0.30	0.15	0.23	0.20	0.3
A2	0.54	0.25	0.25	0.22	0.33
A3	0.99	0.35	0.34	0.29	0.45
B1	0.45	0.20	0.26	0.23	0.35
B2	0.56	0.25	0.26	0.23	0.35
T1	0.38	0.15	0.30	0.26	0.40
M1	1.28	0.40	0.38	0.33	0.50
M2	2.01	0.50	0.45	0.39	0.60
M3	1.61	0.45	0.42	0.36	0.55
E1	0.57	0.20	0.34	0.29	0.45
E2	1.14	0.45	0.30	0.26	0.40
E3	0.96	0.30	0.38	0.33	0.50
C1	0.34	0.15	0.26	0.23	0.35
C2	1.14	0.45	0.30	0.26	0.40
C3	1.59	0.50	0.38	0.33	0.50

Table A3: Random Failures

Non Seismic Random Failures		Human Errors	
RF1	2.50E-03	HER1	2.50E-02
RF2	2.00E-02	HER2	2.50E-02
RF3	7.50E-02	HER3	3.00E-02
RF4	1.50E-02	HER4	5.50E-02
RF5	7.30E-02	HER5	7.50E-02
RF6	2.00E-02		
RF7	5.00E-02		
RF8	3.00E-03		

Tsunami PRA for Kashiwazaki-Kariwa NPP

Keiichiro Saito[a], Masanori Takeuchi[b],Takashi Uemura[c],Yasunori Yamanaka[d]

[a bcd]Tokyo Electric Power Company Inc, Tokyo, Japan

Abstract: The Fukushima Daiichi Nuclear Power Station was struck by the huge tsunami generated by the 2011 off the Pacific Coast of Tohoku Earthquake on March 11, 2011, and experienced a severe accident. The most important lessons learned from the accident was that the "Defense-in-depth for tsunami was insufficient". Therefore we are implementing many safety enhancement measures for tsunami in our Kashiwazaki-Kariwa Nuclear Power Station. We performed tsunami PRA studies in order to evaluate the effectiveness of these measures for addressing tsunami. The studies was based on the guideline "The Standard of Tsunami Probabilistic Risk Assessment (PRA) for nuclear power plants"[1] issued by the Atomic Energy Society of Japan (AESJ) in February 2012. Before and after tsunami countermeasure implementation studies are being done in order to evaluate the effectiveness of the countermeasures. In this paper, the evaluation results for the case of before and after tsunami countermeasure implementation are described, and the effectiveness of the tsunami countermeasures is shown.

Keywords : Tsunami PRA, Kashiwazaki-Kariwa, Fukushima Daiichi accident

1. INTRODUCTION

On March 11, 2011, tsunami generated by the 2011 off the Pacific coast of Tohoku Earthquake hit Fukushima Daiichi Nuclear Power Station (NPS), and it caused flooding at almost all of the seaside area and the surroundings of the major buildings. Then, station blackout (SBO) and loss of ultimate heat sink (LUHS) occurred, and it resulted in severe accidents. One of the lessons leaned by this accident is "Defense-in-depth for tsunami was insufficient". In terms of safety enhancement of nuclear power plant from this lesson, countermeasure for each layer of defense-in-depth against tsunami is enhanced in the Kashiwazaki-Kariwa NPS. In the new nuclear regulation discussed at present, it is required that external event PRA is implemented and existence of sequences other than important sequence groups designated by the Nuclear Regulation Authority is confirmed. Then, we decided to perform tsunami PRA in order to understand plant vulnerability and to check validity of deployed countermeasure against tsunami for Unit 7 (ABWR) of the Kashiwazaki-Kariwa NPS. This paper describes the evaluation result completed by applying to states before and after the implementation of the tsunami countermeasures.

2. OUTLINE OF KASHIWAZAKI-KARIWA NUCLEAR POWER STATION

The Kashiwazaki-Kariwa Nuclear Power Plant(see Fig.1) is located in Kariwa Village and Kashiwazaki city in Niigata Prefecture facing on the coast of the Japan Sea, and seven nuclear reactors (Unit 1-5: BWR5, Unit 6, 7: ABWR, a total of 8,212 MWe) are built.

South　　　　　North

Figure 1. Kashiwazaki-Kariwa NPS

The site is divided into the south side for unit 1-4 and the north side for unit 5-7, and the ground elevation is T.P. 5m (Tokyo Peil: sea-level of Tokyo Bay) at the north side, and T.P. 12m at the south side, respectively. This elevation was decided based on T.P. 3.7m as a result of tsunami height evaluation assumed by wave sources of Echigo-Takada Earthquake from past earthquake record and the related literature. Since then, Tokyo Electric Power Company (TEPCO) reevaluated the tsunami height twice. First time is when the Japan Society of Civil Engineers (JSCE) issued a new design guideline "Tsunami Assessment Method for Nuclear Power Plants in Japan"[2] in 2002. Second time is in 2006 when the latest submarine topographic data and knowledge of fault at ocean areas became available. The results were T.P.3.7m (2002) and T.P.3.3m (2006).

3. TSUNAMI PRA FOR KASHIWAZAKI- KARIWA NUCLEAR POWER STATION

In Japan, from the lesson of the Fukushima Daiichi accident, development of tsunami PRA method was accelerated immediately after the accident, and Atomic Energy Society of Japan (AESJ) issued tsunami PRA guideline in February 2012. Then, TEPCO started to perform tsunami PRA to evaluate the effectiveness of tsunami countermeasures. In the state before the implementation of tsunami countermeasures, since there is no means to prevent flooding to building and function failure of important apparatus assuming generation of tsunami exceeding the 1st floor height of building, each flooding propagation evaluation and fragility evaluation are simply performed, and the core damage frequency (CDF) for each accident sequence is calculated. The items and the contents of the tsunami PRA are described in the following subchapters.

3.1 Tsunami Hazard Evaluation

Tsunami hazard for the Kashiwazaki-Kariwa NPS is evaluated based on "Method of Probabilistic Tsunami Hazard Analysis"[3] issued in 2009 by the JSCE. However, the occurrence frequency and the scale of earthquake, assuming linkage of the multiple faults which is the latest knowledge acquired in the 2011 off the Pacific coast of Tohoku Earthquake, are also taken into consideration.

3.1.1 Tsunami Occurrence Area Model

Regarding the tsunami occurrence area, the tsunami induced by earthquake, originated by faults which exist in the area, is determined in terms of whether they have significant influence on the tsunami hazard of the Kashiwazaki-Kariwa NPS. As a result, the following areas are selected.

1) The fault which is considered in seismic design and is identified by geological survey etc.
2) The fault which is unidentified by investigation, but indicated by an external organization (Epicenter at coast of the Niigata southwest earthquake)
3) The east edge of Japan Sea; Kashiwazaki-Kariwa NPS is considered to be affected significantly when tsunami occurs there.

Regarding these tsunami occurrence areas, the tsunami occurrence scenario is created by setting up the magnitude range and the earthquake occurrence probability.

3.1.2 Uncertainty

Random uncertainty in a numerical computation model and epistemological uncertainty regarding some issues such as existence of active fault and magnitude range etc. are considered in tsunami hazard evaluation. Epistemological uncertainty is dealt with as branch of tsunami occurrence scenario, and given weighting to each scenario. In this evaluation, the magnitude range, earthquake occurrence probability, probability of linkage, and probability distributions of random uncertainty are taken into consideration.

3.1.3 Hazard Curve

The annual probability of exceedance distribution curve is created for each tsunami occurrence scenario defined in chapter 3.1.1 and 3.1.2. Next, for each curve, with consideration for the weighting corresponding to each scenario, statistical processing is performed and hazard curve is created for weighted average as arithmetic average for weighted accumulation sum as fractal curve. As mentioned above, the tsunami hazard curve (tsunami run-up area at the north side) is shown in Fig.2. In evaluation of the state before the implementation of tsunami countermeasures, when tsunami exceeds height of the 1st floor of building, it is simply assumed that flooding in the building occurs and equipment function is lost, and it causes core damage. For example, in the evaluation of Unit 7, since the 1st floor height is T.P.12.3m, when the tsunami beyond this height strikes, it is evaluated as core damage occurs.

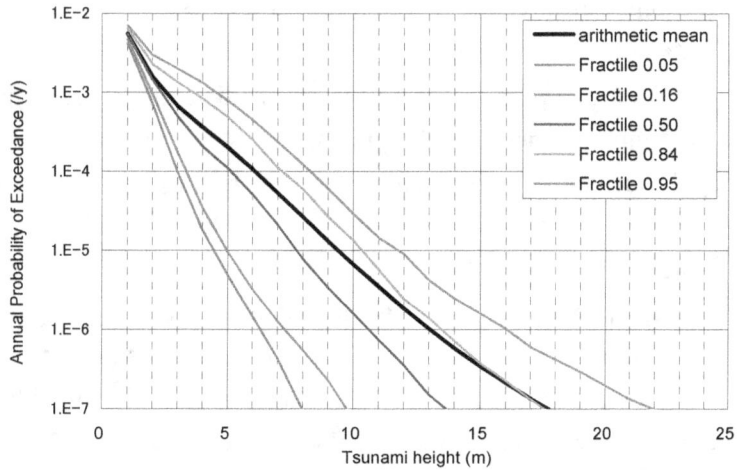

Figure 2. Tsunami Hazard Curve

3.2 Tsunami Fragility Evaluation

Regarding influence to apparatus by tsunami, damage by flooding and by tsunami wave force is considered. Regarding equipment on yard and door on outer wall of buildings such as yard tank, yard watertight door, etc., the failure probability against tsunami wave force is set by flooding depth based on tsunami run-up analysis result. Regarding equipment and door inside building, the damage probability is set by flooding propagation analysis result for building. Regarding tsunami run-up analysis, it is performed for multi case of tsunami height. For each case, fragility curve is evaluated from the equipment damage probability with consideration for the uncertainty in the flooding depth of the installation location for each equipment. The views of the main objects are shown below.

1) Embankment, tidal wall

When tsunami exceeds the height of the embankment or tidal wall, these failures are assumed.

On the other hand, in case of the tsunami less than the height, since there is sufficient resistance stress in the design, it is hard to consider to be damaged. For the reason, the height of embankment and tidal wall is set as failure-of-function limit.

2) Watertight door, general door

Regarding protection doors installed on building outer wall, fragility evaluation is conservatively performed with consideration for tsunami wave force. On the other hand, it is assumed that flow velocity of flooding propagation in building is slow enough, and fragility evaluation for the door in building is conducted for hydrostatic pressure of water accumulated in division.

3) Yard tanks (Light oil tank, pure water storage tank)

Since these tanks are on the ground, damage evaluation by tsunami wave force is performed, but evaluation for flooding and function affected by water level by submersion is also performed.

4) Fire protection system piping

Fracture evaluation is performed for bending load of piping changed by tsunami wave force. Branch piping which has high failure possibility is also taken into consideration.

5) Equipment in building (reactor core isolation cooling system (RCIC), power panel, etc.)

Flooding propagation evaluation in building is performed, and when the concerned apparatus and required support system are inundated, the function failures are assumed.

However, in evaluation of the state before the implementation of tsunami countermeasures, fragility evaluation with consideration for uncertainty is not performed, but method that the events induced by the tsunami of a certain height are deterministically evaluated is adopted.

3.3. Accident Scenario Identification

3.3.1 The state before the implementation of tsunami countermeasures

At the state before the implementation of tsunami countermeasures, it is assumed accident scenarios considering flooding according to the tsunami wave height. In addition, if the tsunami height is below the site level (T.P. 12m), it is assumed that inundation starts via maintenance hatch (T.P. 3.5m) in the heat exchanger area in the turbine building when tsunami height exceeds T.P. 3.5m. Also, it is conservatively assumed that all the buildings connected to turbine building are flooded to the tsunami height.

Tsunami height between T.P. 4.2m and T.P. 4.8m

The support system (ex. reactor cooling water system (RCW) pumps, reactor sea water system (RSW) pumps) is located in basement 1st floor of turbine building (T/B). When tsunami height exceeds T.P. 4.2m, the support system is flooded and it causes LUHS by the function failure. In addition, non-safety related metal-clad switch gear (M/C) in basement 2nd floor of T/B is also flooded.

1) Tsunami height between T.P. 4.8m and T.P. 6.5m

Emergency M/C in basement floor of reactor building (R/B) is flooded and lost its function. It causes SBO by the function failure of emergency M/C and non-safety related M/C, because it cannot be powered by off-site power and emergency diesel generators (D/Gs).

2) Tsunami height between T.P. 6.5m and T.P. 12.3m

DC power panel in the basement floor of control building (C/B) is flooded and loses its function. It causes loss of DC power.

3) Tsunami height exceeding T.P. 12.3m

Tsunami runs up to the site level, low voltage start-up transformer located at the site level is flooded and loses its function, and inundation into the main buildings occurs via entrance of each building.

3.3.2 The state after the implementation of tsunami measures

Using the results of tsunami fragility analysis as a reference, initiating events which are induced by tsunami are adopted and accident scenario analysis is conducted.

The extracted initiating events are shown below,

1) Loss of off-site power (LOPA)

    # Flooding of low voltage start-up transformer

2)   Loss of function of emergency D/G

    # Flooding of emergency D/G(A,B,C) by inundation of R/B

    # Fuel transport failure by damage of light oil tank

    # Fuel transport failure by damage of fuel transport pump

    # Operation failure of emergency D/Gs operation failure by loss of support system function by T/B flooding

    # Flooding of emergency power panel room in R/B

3) Loss of ultimate heat sink

    # Loss of support system function by T/B flooding

    # Loss of support system function by D/G failure (in case of LOPA)

4) Loss of instrumentation and control system function

    # Flooding of main control room (MCR) in C/B

    # Flooding of DC power panel in C/B

3.4 Accident Sequence Evaluation

The evaluation result of the state before and after the implementation of tsunami countermeasures are described below.

3.4.1 The state before the implementation of tsunami countermeasures

Accident scenario changes according to tsunami height. So, initiating events and credited mitigation systems are changed as well.

1) Tsunami height between T.P. 4.2m and T.P. 4.8m

Initiating event is set as LUHS. In identified accident scenario, the relief valve function of SRV and RCIC are credited as mitigation systems. Event tree is shown in Figure 3. CDF for this tsunami height is calculated as 8.8E-5(/RY) and dominant sequence is TQUV (Transient with loss of all ECCS injections).

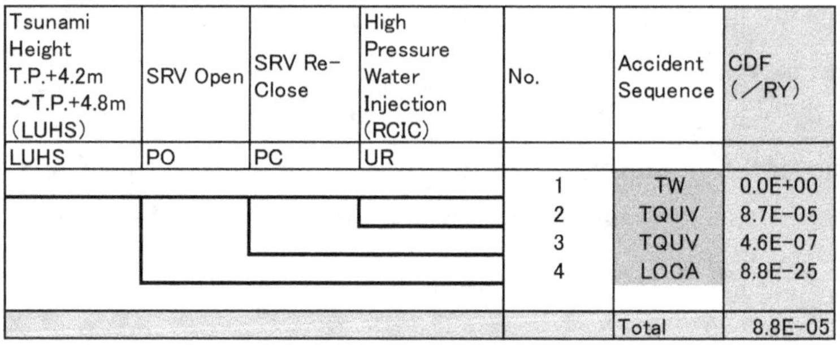

Tsunami Height T.P.+4.2m ~T.P.+4.8m (LUHS)	SRV Open	SRV Re-Close	High Pressure Water Injection (RCIC)	No.	Accident Sequence	CDF (/RY)
LUHS	PO	PC	UR			
				1	TW	0.0E+00
				2	TQUV	8.7E-05
				3	TQUV	4.6E-07
				4	LOCA	8.8E-25
					Total	8.8E-05

Figure 3. Event Tree (Tsunami Height T.P. 4.2m~4.8m)

2) Tsunami height between T.P. 4.8m and T.P. 6.5m

Initiating event is set as LUHS and SBO. Credited mitigation system is same as 1). Event tree is shown in Figure 4. CDF for this tsunami height is calculated as 1.0E-4(/RY) and dominant sequence is TQUV.

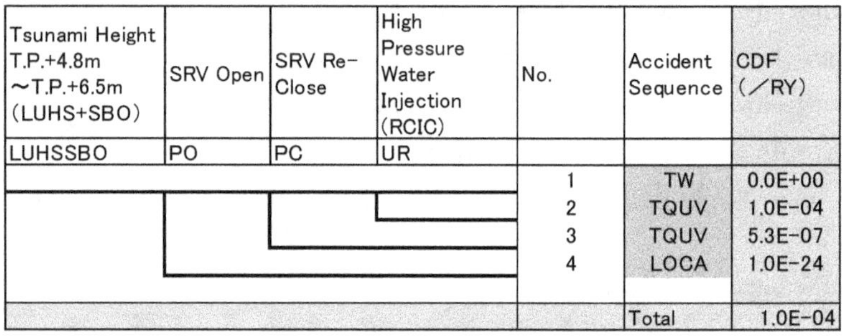

Tsunami Height T.P.+4.8m ~T.P.+6.5m (LUHS+SBO)	SRV Open	SRV Re-Close	High Pressure Water Injection (RCIC)	No.	Accident Sequence	CDF (/RY)
LUHSSBO	PO	PC	UR			
				1	TW	0.0E+00
				2	TQUV	1.0E-04
				3	TQUV	5.3E-07
				4	LOCA	1.0E-24
					Total	1.0E-04

Figure 4. Event Tree (Tsunami Height T.P. 4.8m~6.5m)

3) Tsunami height exceeding T.P. 6.5m

Initiating event is set as LUHS, SBO and loss of DC power. No credited mitigation system is set because it is assumed loss of DC power. Event tree is shown in Figure 5. CDF for this tsunami height is calculated as 2.5E-5 (/RY) and dominant sequence is TBD(Transient with loss of all AC & DC powers).

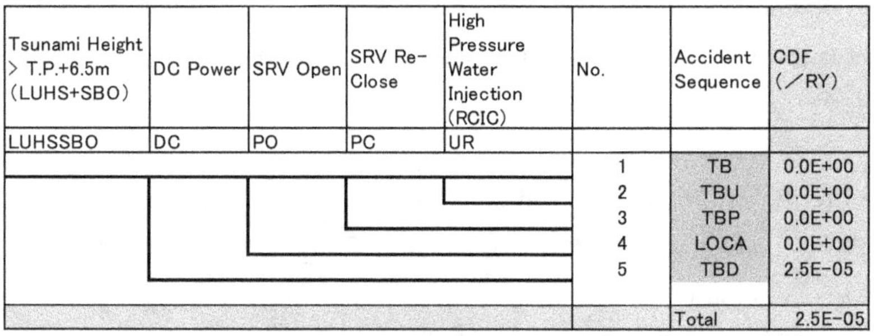

Tsunami Height > T.P.+6.5m (LUHS+SBO)	DC Power	SRV Open	SRV Re-Close	High Pressure Water Injection (RCIC)	No.	Accident Sequence	CDF (/RY)
LUHSSBO	DC	PO	PC	UR			
					1	TB	0.0E+00
					2	TBU	0.0E+00
					3	TBP	0.0E+00
					4	LOCA	0.0E+00
					5	TBD	2.5E-05
						Total	2.5E-05

Figure 5. Event Tree (Tsunami Height >T.P. 6.5m)

Tsunami PRA result at the state before the implementation of countermeasures is shown in Figure 6. Total CDF is calculated as 2.1E-4(/RY) in average value. As for accident sequence rate, TQUV is dominant sequence accounting for 89 percentages

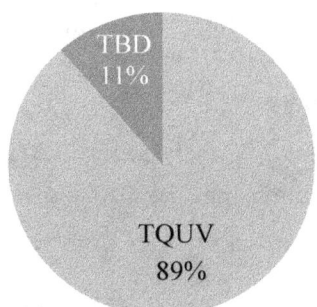

Figure 6. Tsunami PRA result
(The state before the implementation of tsunami countermeasures)

3.4.2 The state after the implementation of tsunami countermeasures

Based on the result of tsunami fragility analysis, in the accident sequence analysis, failure rate for each system, structure and component which is relevant to initiating events or equipment relevant to credited mitigation system is calculated and combination of tsunami height and damaged equipments is considered.

Regarding to the accident sequence analysis, tsunami initiating hierarchy event tree is constructed. In this event tree, yard equipments whose failure are directly connected to the initiating event are set as headings. The hierarchy event tree is shown in Figure 7. In event tree for each initiating event which is expanded from the hierarchy event tree, yard equipments which are not considered as heading is set as mitigation systems.

Tsunami	Off-site Power (Embankment)	DC Power	AC Power+RCIC	Support System	AC Power (Emergency DG)	Accident Sequence	CDF (/RY)
TU	BO	DC	RC	RW	AC		
						–	
	TS_BOUCHO.ft					To LOPA	–
					TS_AC-480CDE.ft	To SBO1	–
				TS_TB_FLOOD.ft		To SBO2	–
			TS_RB_FLOOD.ft			TBU	2.4E-08
		TS_CB_FLOOD.ft				TBD	7.6E-08
						Total	1.0E-07

Figure7. Hierarchy Event Tree

The outline of accident sequence analysis is described below.

1) Tsunami height between T.P. 15m and T.P. 17m

Because, as shown by the fragility analysis result, the water tight doors of each building are not broken by tsunami of this height, inundation into the buildings does not occur, but the fuel transport pumps on yard are destroyed by tsunami. In this state, random failure of temporary oil transport pump which is installed thereafter is assumed. Because of this, all emergency D/Gs lose their function, and it causes the SBO.

2) Tsunami height between T.P. 17m and T.P. 18m

Because, as shown by the fragility analysis result, the water tight doors of T/B and R/B are broken by tsunami of this height, inundation into the T/B and R/B occur. Inundation into the T/B causes the flooding of support systems (ex. RCW and RSW pumps) and the loss of its function, and then LUHS occurs. Also, inundation into the R/B causes the flooding of RCIC control panel and the loss of RCIC function. Then all of the water injection function failure is occurred.

3) Tsunami height exceeding T.P. 18m

Because, as shown by the fragility analysis result, the water tight door of C/B is broken by tsunami of this height, inundation into the C/B occur, and it causes the loss of DC power (TBD).

Tsunami PRA result at the sate after the implementation of countermeasures is shown in Figure 8. Total CDF is calculated as 1.0E-7(/RY) in average value. As for accident sequence rate, TBD is dominant sequence accounting for 74 percentages in total CDF.

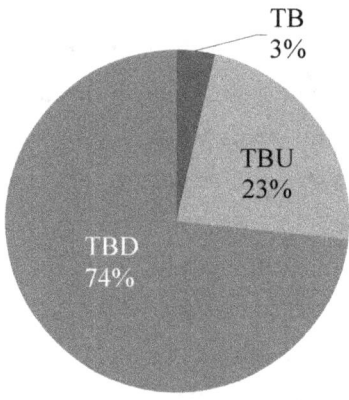

Figure 8. Tsunami PRA result
(The state after the implementation of tsunami countermeasures)

4.EFFECTIVENESS EVALUATION ABOUT THE MEASURE TAKEN IN THE KASHIWAZAKI-KARIWA NUCLEAR POWER PLANT

As stated above, in Kashiwazaki-Kariwa NPS, the various safety countermeasures is being deployed reflecting the lessons learned from the Fukushima Daiichi NPS accident. The measures against tsunami and power supply are included in the countermeasures, and the validity of these measures will be evaluated by using the tsunami PRA. Here, the validity for the implemented safety measures is qualitatively discussed from the view of TQUV and TBD which are the important accident sequences determined prior to the implementation of additional safety countermeasures. Regarding TQUV, probability of LUHS and possibility of inoperable of RCIC by submersion will decrease due to installation of embankment, tidal wall and watertight doors for important equipment rooms such as RCIC room and modification for maintenance hatch in T/B. Furthermore, when all low pressure water injection systems loses by tsunami exceeding the embankment height, water injection can be done by fire engines located at high elevations. Therefore, in the state after the implementation of the tsunami countermeasures, it can be presumed that the occurrence probability of TQUV is reduced substantially. As for TBD, probability of LOPA and inoperable possibility of DC power by submersion will also decrease due to installation of embankment and watertight doors of important equipment rooms. In addition, the enhancement of DC power supplies is implemented for storage battery extension at higher floor in the reactor building, additional established storage battery, installation of the small generator, and maintenance of the DC power supply means. Accordingly, it is presumed that the possibility of loss of DC power decreases. Therefore, the present measures can be presumed as being appropriate against the important accident sequences extracted.

5. CONCLUSION

Tsunami PRA studies for Unit 7 of Kashiwazaki-Kariwa NPS was conducted and the dominant accident scenarios that may result in core damage due to flooding were identified. The important accident sequences were evaluated as TQUV and TBD at the state before the implementation of countermeasures and CDF calculated as 2.1E-4 (/RY). This information supports qualitative assessment of the countermeasures that have been and will be implemented which indicates that these accident sequence probabilities will be decreased. Hence, the tsunami PRA was performed with the state after the implementation of tsunami countermeasures and CDF is calculated as 1.0E-7(/RY).

By comparing these two CDFs, the effectiveness of the tsunami countermeasures which are implemented in the Kashiwazaki-Kariwa NPS is confirmed. TEPCO will continue to evaluate the risks of external events including tsunami using PRA methods and enhance safety of the Kashiwazaki Kariwa NPS using such results.

References

[1] "The standard of Tsunami Probabilistic Risk Assessment (PRA) for nuclear power plants: 2011", AESJ, February 2012

[2] "Tsunami Assessment Method for Nuclear Power Plants in Japan", JSCE, February 2002

[3] "Method of Probabilistic Tsunami Hazard Analysis", JSCE, March 2009

Reliability and Safety Models of Transportation Systems - a Literature Review

Franciszek J. Restel[*]

Wroclaw University of Technology, Wroclaw, Poland

Abstract: Transportation systems form the backbone of economy and play an important role in society. Because of the far-reaching effects of disruptions on these systems (social, economic, national defense), they are classified as critical infrastructure systems.

Reliability researches on various elements of transportation systems were carried out since the mid-twentieth century. The focus was on vehicles and their components.

Infrastructure is an important component of system, in addition to vehicles. The highest level of complexity is characterized by the railway infrastructure. It is natural, that this led to a number of models describing selected issues.

In recent years, much attention has been paid to critical infrastructure systems. There have been numerous proposals for the use of graph models in the analysis of resilience and vulnerability of transportation systems.

There are many groups of models describing reliability. Some models contain reliability factors only fragmentary.

This paper presents experience with reliability models, were tested in research work on the railway transportation system. The inference is not limited to railway transportation system, but generally relates to land transportation systems.

The review includes also own models that are dedicated to describing reliability of fixed-track systems, in which processes are determined by a timetable.

Keywords: Railway, Transportation Systems, Reliability, Safety, Models.

1. INTRODUCTION

There is a large ambiguity in nomenclature in the field of the rail transportation system as far as the connecting areas of such terms as reliability, transportation systems and railway engineering are concerned. The most common examples of inconsistencies cab be found in the studies of transportation systems that discuss elements of reliability. This problem has been revealed after a preliminary literature review.

Public transport companies show a common tendency to narrow the facets of reliability down to proper execution of transportation tasks - i.e. punctuality [58]. Studies of this aspect are limited to modelling of arrival and departure times [4,58]. Furthermore, punctuality is also the sole aspect mentioned in papers presenting models used for selection of transportation modes in public transport services [5]. The term "travel time reliability" is introduced, denoting punctuality of inter-connected services at transfer nodes. Travel time reliability is, first and foremost, mentioned in analyses that refer to passengers [40]. In [44] travel time reliability is defined as probability of travelling a certain route in a period of time that is shorter than or equal to the assumed travel time.

Most often, however, reliability is understood as punctuality, i.e. an attribute referring to process execution in accordance with the established schedules. Chen [12] enumerates the most commonly appearing terms in the context of reliability in the railway system, by listing the following definitions:
- Reliability - probability of performing a required function by an object under given conditions and in a specified period of time,

---

[*] franciszek.restel@pwr.edu.pl

- Railway transportation services reliability - participation of trains actually starting at a given time (including reserve trains) in the total number of all scheduled trains in a theoretical timetable. This attribute does not take into account the size of (original and secondary) disruptions,
- Punctuality - participation of timely train arrivals at every station in all arrivals scheduled in a timetable.

Punctuality is considered one of the most important measures of reliability of transportation system processes [28,47,49,67]. In practice, punctuality is not measured for every train arrival and departure to and from every single station, but it is only verified at the destination station of a given run [49].

Vromans [66], however, gives the following measures, as the most popular aspects in determining reliability of a rail transportation system:
- punctuality,
- punctuality of inter-connected services - concerns delays resulting from the loss of inter-connected train service,
- number of cancelled trains,
- average train delay,
- average passenger delay,

Factors that influence punctuality of trains in the rail transportation system were identified in [49,58]:
- number of passengers,
- degree of vehicle occupation in a train,
- use of traffic capacity,
- cancelled services,
- temporary speed limits,
- technical support of the infrastructure,
- train traffic organization.

The authors in [4] take into account specific ranges of delay in punctuality analyses. They quote two minutes as the border value. However, when presenting the results obtained from the study of an actual system, the border value was established as 5 minutes. Some scientific studies also propose the border value as 2.5-minute delays [48]. However, such values refer to rail transportation systems in highly developed societies (e.g. Norway). Such rail transportation systems have become more efficient as a result of significant society development. Along with the development of societies, new types of constraints are put on antropotechnical systems. One such example are Scandinavian railway services with high punctuality standards and their adverse events research focusing, in fact, on the cases of suicides [53,54]. In [65] Vansteenwegen defines punctuality in the context of a train's arrival at the destination station. At the same time it is stated that delays up to 5 minutes are a good result of system operation.

2. RELIABILITY IN RAIL TRANSPORTATION SYSTEM

From the very beginning of reliability studies in rail transportation system studies mainly focused on vehicles. Apart from testing mathematical models, operational data analyses were conducted concerning damage. The research was narrowed down to a statistical analysis and conclusions drawn therefrom. Rail vehicles consist of systems, assemblies, subassemblies, etc. which may be subject to failure resulting in the entire vehicle becoming non-operational. Thus when analyzing their reliability decomposition is performed in order to design the models more precisely [59]. Operational studies into vehicle failure rates are used, among others, to determine optimal inspection and repair intervals.

A crucial aspect in the railway engineering practice is determination of the state of infrastructure and relating safety improving actions to it. [3]. From such a perspective introduction of train speed limit is considered an alternative to technical service. Both the choice of when to introduce a speed limit and performance of technical service entail costs. A crucial group of studies in this aspect involve the so-called Life Cycle Costs analyses [41]. In [23,29] decision models for determining the right time for technical service of infrastructure at minimum total costs have been designed.

The above-mentioned models take into account the costs of a planned technical service, speed limitation costs, however, they do not include any aspects of reliability. The costs related to adverse events are not considered. Another group of tasks performed in rail transportation system includes a dispatcher's actions once disruptions occur. One of the possible methods is to regulate the speed by running trains in order to minimize unplanned stopovers (particularly perceived by passengers) and reduce energy consumption [16]. In this approach no structural process changes are proposed, but merely changes in process parameters. In fact, these aspects are related to re-organization of traffic after occurrence of disruptions whose aim is to minimize further propagation of disruptions [10,62,63].

A more detailed insight is provided by an analysis of consequences of original and the related secondary damage (excluding the traffic impact) [60]. Chen in [12] presented train services reliability and punctuality models. The train service reliability model is composed of three sections:
- reliability of technical facilities constituting the system (independence of events occurring in subsystems was assumed),
- interactions between the transport modes subsystem and other subsystems,
- intensity of restoring the traffic after interruption caused by an adverse event.

The authors [33] proposed a disrupted train traffic management support model. The model is based on the costs of rail traffic reorganization or cancellation of trains in the context of railway employees (engine drivers and train traffic service staff).

The probability of delay suppression is directly related to the so-called "resistant timetables" [36] and resilience [22], i.e. an ability of the system to regain functionality after an event. Vromans [66] narrows the term of rail transportation system reliability down to reliability of transport services. Time disruptions constitute the only aspect that is taken into account. The author identifies potential causes of disruptions, however, he assumes that the time disruptions have only one source. The paper focuses on max-plus algebra used for assessment of timetables (cf. [27,28]) in the context of time reserve in a timetable. A support model for decision-making during traffic control in the case of disruptions was presented in [1]. In fact, a crucial impact of infrastructure on propagation of disruptions was noticed. Decision variables include the moments when a k-event begins and finishes.

The previous groups show a tendency to narrow the subject down to one specific aspect. The next group of aspects has a completely different approach in comparison to the before-mentioned ones. They include research in which the impact of catastrophic events on system operation is analyzed. Railway services are in this respect considered Critical Infrastructure System (CIS), whereas conducted analyses focus on serious events with dire consequences. A CIS description contains graph models in which the most basic ones are modelling simply the relations between junctions. The more advanced models include traffic capacity of the edges, travel time, traffic control at junctions, as well as the mode of power supply [18,19]. In this aspect the term - reliability of railway network infrastructure traffic capacity - is introduced [72].

3. FUNCTIONAL RELIABILITY MODELS

In [71] Zamojski presents a functional reliability model of a discrete transportation system. A discrete transportation system is further defined as a relation in the Cartesian product of transportation task sets (theoretical and actually performed), means of transport, infrastructure, a dispatcher and time. The described time is the so-called network chronicle [71]. A resource set called functional system configuration is assigned for the purpose of input task execution. An input task is defined as a priori by a dispatcher based on available resources. An output task is the actual execution of a transportation task. Reliability characteristics are attributed solely to vehicles.

In [20] a functional reliability model of track availability control system was presented. The model is based on the following three layers:
- – a traffic process along the track system,
- – an operational process of train traffic control devices,
- – a decision-making process in management of train traffic control devices.

Layer 1 concerns arrangement of tracks and vehicle flow with possible disruptions. The second layer (operational process layer) ensures ordering of routes into admissible and inadmissible at a given moment. It defines whether two train courses can be run at the same time without a risk of collision. The third layer is based on a decision model generating a series of decisions that control the processes based on available information from the two remaining layers. Three types of adverse events related to availability-identifying devices have been distinguished:
- – "catastrophic" (accidental) damage causing irreversible changes in the attributes of the system, whose removal is possible only by replacing technical elements,
- – disappearing damage, resulting from a temporary overrun of work parameters can be "cancelled" by an employee,
- – wear-out damage, whose removal is possible only through replacement of technical elements.

In [51] a railway network simulation model was presented. Two railway lines with a common middle section were divided into equal sections. Every section can assigned one of the six states of degradation, where *0* is the incapacity state. The intensity values describing transitions between the states are constant. For the purposes of the model, traffic volume and time spent by a train for travelling each section were defined. The degradation model has been completed by possible line speed values (nominal speed 100 km/h, reduced speed - 80 or 60 km/h) by obtaining an 18-state model allowing for determination of a daily delay on a rail network.

4. MODELS BASED ON STATE-TRANSITION GRAPHS

One of the first reliability appraisals of rail vehicle elements were based on Markov processes, a reliability structure analysis and a fault tree. In the recent years Markov processes have been also applied in modelling of degradation and critical damage of the track structure [17]. The authors in [11] discuss an analysis of reliability and safety of rail traffic control device components in the aspect of safe failures. The paper focuses on an analysis of a junction reliability and safety analysis involving special microcomputers used for calculations, called transputers. To this end, a 9-state Markov model was developed. The model does not present the state of safety faults. The states with unrevealed errors or errors leading to similar output signals (potentially erroneous) form security threats.

In [70] a functional and reliability model of a discrete transportation system was presented. The model, similarly to the previous functional and reliability models narrows down in the aspect of reliability to a vehicle subsystem, since Markov processes have been used in it solely to describe reliability of vehicles. A series of assumptions regulating the functional part has been assumed in the model. The reliability part has been described by the Markov model in which states are defined by the number of damaged vehicles. Transportation system has been described by a matrix differential equation for n-recipient:

$$\frac{d}{dt}P^{(n)}(t) = P^{(n)}(t)\varrho^{(n)}$$

where:
$P^{(n)}(t)$ - state probability vector,
$\varrho^{(n)}$ - transition intensity matrix.

The authors [14] presented a Markov model of tram reliability systems. Six possible reliability states have been assumed for vehicles:
- – *TTB* – availability state,
- – $\bar{T}TP$ – partial availability due to damage to one out of two driving modules,

- $TT\bar{B}$ – vehicle failure due to damage to the braking module,
- $\bar{T}\bar{T}B$ – vehicle failure due to damage to two drive modules,
- $\bar{T}T\bar{B}$ – vehicle failure state due to damage to one drive module and a breaking module,
- $\bar{T}\bar{T}\bar{B}$ – vehicle failure state due to non-serviceability of two drive modules and a braking module.

The authors emphasized that the time required for a tram repair has a non-exponential distribution and assumed Erlang-3 distribution. Due to the connections with exponential distribution it is possible to represent Erlang-3 distribution in Markov model. One repair state with Erlang-3 distribution is replaced by three states described by the same exponential distributions. Due to the characteristics of the actual system it was assumed that at a given time, there can be maximum three fully or partially damaged vehicles. It was shown that for the analyzed tram transportation system the use of exponential distributions for a description of repair times shows higher availability of the system than in reality. The use of approximation of Erlang-3 distribution allowed for a more detailed modelling of the system, more convergent with actual results.

In [9] the Markov model was used to analyze the impact of various instances of faults on the availability of the railway transportation system. A four-state model in which the following states were specified has been proposed:
- (1) - availability state,
- (2) - safe failures occurred in the system which do not hold train traffic.
- (3) – hazardous errors occurred in the system - safety threat state with running train traffic,
- (4) – train traffic stoppage in the system in order to remove the threat.

Introduction of state (2) was caused by continuation of traffic in the actual system after occurrence of safe damages. Observations of the railway transportation system show that assessment whether damage is safe or whether it depends on the traffic supervisor. Thus introduction of state (3) is justified. The authors of the model did not predict the state of safety loss, by introducing interchangeable a threat removal state (4).

A frequent subject of tests is an element of railway transportation system. The example can be analyses of technical facilities belonging to the infrastructure. In [32] a 40-state model of reliability and safety of the railway crossing security equipment. Apart from intensity of damage and repairs, the model also includes detection of failures which was the reverse of the time between inspections. The first model was simplified to 21 states.

In [47] Markov processes were used to analyze reliability and safety of the combined transportation system (road-rail). In [46] a three-phase Markov model of intermodal transportation system was presented. The phase structures are identical. Introduction of phases results from the change of transportation mode in the case of intermodal transportation. The three phases represent road, rail and road transport. The author used the model for determination of availability function for intermodal transportation depending on the time proportion of individual phases. By analogy, Markov processes were applied for a multimodal transportation system [34]. The authors of the paper introduce the term of functional reliability defined as probability of supplying the right amount of load in the time not exceeding the exponential time.

Other areas in which Markov processes are used include:
- a train aggregation analysis in a selected subsystem [37],
- a reliability analysis of rail vehicles with their operational systems.

Markov chains have found application e.g. in:
- Bayesian analysis of collective bus transport routing [35],
- modelling of toll collection systems [68],
- technical service scheduling with the use of fault trees [45],
- in modelling of an aging system, including correction services [38].

5. SAFETY IN TRANSPORTATION SYSTEMS

Safety is related to maintenance of the system state that prevents occurrence of adverse events, such as [43]:
- death,
- body injuries,
- tangible property loss,
- natural environment loss.

In [61] a method of barrier identification based on the fault tree was presented. The method is based on the so-called Swiss cheese model, in which the holes must overlap so that an arrow can go through them (for safety failure to take place). A risk situation, which is a peak event is modelled by a classic fault tree with AND and OR logic gates. Once the tree is drawn up the first logic gates are searched for starting from the peak event. The event above a given OR gate is directly related to one barrier and used for determination of a barrier.

In the case of railway transportation system an event tree and a fault tree can be used in adverse event risk analyses. In [2] such an analyses was expanded by addition of risk influencing factors. The problem was shown on an example of a single-track line, for which a peak event was a collision of two trains coming from two different directions. Barriers which aimed at preventing occurrence of peak events were catalogued and then a tree of events leading to the barrier faults were drawn up. Operational risk influencing factors were attributed to the basic events.

The studies [6,7,56] explore security engineering in rail transportation system design. Articles discuss the issue of ERTMS (European Rail Traffic Management System) implementation, which in their structure also contain a unified European communications standard GSM-Rail. The problem of security at ERTMS implementation is all the more crucial if we take into account lack of experience in operation of such a system in conditions corresponding to the implementation (for Dutch railways in 2003). The basis for discussions [6] is risk identification for the implemented system. The [7] publication summarizes literature which introduces risk analysis components for a newly designed railway system (ERTMS). For security appraisal of the ETCS system, a slightly simplified ERTMS variant, Functional Hazard Assessment method was proposed in [55].

A crucial aspect in the assessment is identification of Safety Integrity Levels (SIL). In the draft of the European standard [24] a simplified SIL table was used for railway traffic control, communication, data processing equipment and electronic systems significantly influencing the safety of rail transportation. In [8] representing railway transportation risk levels in the form of a table was suggested (Table 1). In this way unacceptable risk, acceptable risk areas and a border area were obtained. Combinations of frequency and consequences in borderline risk areas were then used for drawing up a risk table based on the SIL table. The proposed table presents frequency of event occurrence and consequences divided into A to E. Group A represents events which can be classified as fail-safe. The remaining groups have been divided in terms of energy accumulated during the event:
- B - concerns consequences of events during shunting,
- C - concerns events at low linear speed values,
- D - events at medium linear speed values,
- E - concerns consequences of events at high speed values.

Table 1: Permissible Risk Levels Table. Prepared on the Basis of [8]

Failure rate, per hour of use	A	B	C	D	E
$> 10^{-5}$					
10^{-5}	acceptable			not acceptable	
$3 \cdot 10^{-7}$					
10^{-8}				border	
10^{-9}					

In [15] risk assessment of hazardous load transportation was conducted by using events/vehicle-kilometers as a measure of event occurrence intensity. In [39] a model used in a risk analysis of hazardous material rail transportation was presented. The risk is determined as the product of intensity of derailment of carriages used for transport of hazardous materials, operational work related to transportation of hazardous materials, conditional probability of hazardous material release after derailment and consequences of hazardous material release from the carriage.

6. HUMAN FACTORS

The literary sources point out that the human factor dominates during the occurrence of hazards [30,64]. In the case of standard rail traffic control devices a situation may occur in which the security system will have to be circumvented to enable further operation of train traffic. Emergency traffic operation is conditioned by improper system operation which constitutes a possibility for unreliability of security [21]. In [26] a model for human error occurrence probability during operation of rail traffic security system was presented:

$$HEP = HEPgen \cdot [(EPC - 1) \cdot EFF + 1]$$

where:
HEP - human error probability,
$HEPgen$ - intensity of human error probability in atypical situations,
EPC - coefficient describing whether work conditions can contribute to a human error,
EFF - coefficient describing efficiency of error-conducive conditions in error occurrence.

Increased behavioral and cognitive load has an impact on traffic safety (more than 90% of accidents in the rail transportation system occur after taking over the responsibility by a human [56], whereas disasters in transport occur in around 80% of cases due to a human error [31]. A human factor exists in the entire system not only in direct operation of trains by railway traffic control stations [69].

A weighty problem in the use of the rail transportation system is the SPAD (Signal Passed At Danger) phenomenon. The most common cause of this type of events is the machine operator's error. The [25] source says that in 2011 more than 45% of accidents were caused by a train passing a stop signal in a dangerous way. SPAD events were examined qualitatively in [50] by using the so-called SHEL model SHEL is an acronym which stands for Software-Hardware-Environment-Liveware. The software is understood as a collection of principles, guidelines of operation and other practices which define cooperation of the system elements. The hardware represents all technical facilities belonging to the system. The environment represents social, political and economic influence on the system from the outside. The livewear represents a collection of people in the system and their influence on the executed processes.

Due to the crucial impact of those events on occurrence of safety failure they are examined in detail by using various methods (e.g. Bayesian networks [42]).
In the context of a human factor a science dealing with safety culture in antropotechnical systems should be noted. Models of the problems were synthetically introduced in [13].

7. SPATIAL REPRESENTATION OF THE RAIL TRANSPORTATION SYSTEM STATES

A model of rail transportation system can be presented as a process described by states, which is divided into a specified number of subsets. The most commonly distinguished subset is the availability and failure subset. Defining of states from the rail transportation system perspective is a complex problem therefore they were divided into constituent parts. When analyzing reliability of the system, its operational availability (in execution of transportation tasks) should be the first issue examined. In this aspect the set of states has been divided into subsets:
  - availability of the system in execution of transportation tasks,

- partial availability of the system in execution of transportation tasks,
- unserviceability of the system in performance of transport tasks.

Consequences of incorrect operation are introduced in the second dimension:
- none,
- small,
- big.

In the third dimension organizational consequences of unwanted events (delays) are introduced:
- none,
- small,
- big.

The possibility of unwanted event occurrence and spreading of the consequences depends on the intensity of the system operation expressed as operational work [57] or transportation work (transportation load) [52]. As a result a super cube state defining model was developed, for state defining in four dimensions (Figure 1 and Figure 2).

Figure 1: State defining cube – three dimensions

Figure 2: Fourth dimension of the super cube

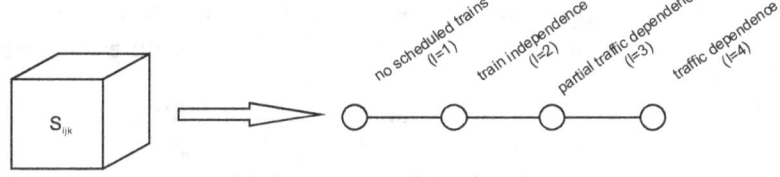

8. CONCLUSION

Research on the reliability and safety of rail transportation system can be divided into the following groups of models:
- Safety in rail transport:
    - risk models,
    - models describing the human factor,
    - models of the formation of accidents (including the failure of barriers)

- Reliability models based on ST graphs:
    - system models taking into account the dangerous failures,
    - system models taking into account only secure failures,
    - models of selected subsystems,
- Functional-reliability models:
    - system models with breakable infrastructure,
    - system models with breakable fleet,
    - models of breakable subsystem,
- Reliability models of transport processes:
    - models with random streams of applications and operating,
    - models with random unwanted events,
    - models with random travel time,
    - models with unwanted events occurring according to scenario.

Risk models are used to assess the occurrence and consequences of unwanted events. Models used for transport systems take into account the intensity of use of the system (volume of traffic), the probability of system corruption, improper use of the system (for example, against the technical rules of conduct) and human factors. Models describing human factors determine the unreliability of man in different circumstances the systems condition. The improper use of the system (conscious or unconscious errors) is taking into account and the process which will lead to the loss of security is examined. Models of accidents and analysis of the barriers, include physical dependence of trains movement, improper use of the system, human factors and risks. The process which leads to that failure of safety is determined.

Reliability models of rail transport system based on state-transition graphs (ST) with dangerous failures, include the failure of rolling stock or the general unreliability of the system. States of improper use of the system and the unreliability of security are modelled. Another group are functional and reliability models. These are primarily functional models of systems with selected reliability characteristics. Calculations of such models are most often performed using computer simulation. Analyzed system models take into account the failure of the infrastructure or of the rolling stock.

The next group are models of reliability of transport processes. Models of mass service system include random streams of applications and service. The unreliability of the system is contained in the random streams. Models with random driving time also apply to capacity and punctuality. The unreliability of the system is included in the randomness of driving time. In these models, interference in the process of transport are taken into account only in the form of delays.

Models of unwanted events according to the scenario relate primarily to examine the effects depending on the event. The effects are most often reported to the functions of the cost. These models are related to subsystems cycle cost analysis, using time reserves, losses due to interference and reorganize the movement. A common feature is skipping the mishap. In the other hand, the traffic and the intensity of use of the system are taken into account. Most models of critical infrastructures fall into this group.

In conclusion it should be noted that in previous studies of the reliability and safety of rail transport system developed models that consider individual characteristics. Security models do not take into account such features as motor disturbances or physical dependence of trains movement. Reliability models based on graphs ST do not include those features. In addition, limited to include only damages the individual subsystems or the system as a black box. Inventoried models of functional reliability show all functional aspects, taking into account only unreliability of one subsystem or component. Reliability models of the transport process focus on the assessment of the timetable, however, aspects of security and misuse are ignored.

Due to the nature of the rail transport system and the lack of appropriate models, it was developed a model of reliability and safety, which include the following features:
- The intensity of use of the system,

- Dependence on trains movement,
- The failure of the infrastructure and rolling stock,
- Providing disrupted the transport process,
- Improper use of the system,
- The human factor,
- Threats and security failure, process which will lead to loss of security.

References

[1] R. Acuna-Agost, P. Michelon, D. Feillet and S. Gueye. *"SAPI: Statistical Analysis of Propagation of Incidents. Anew approach for rescheduling trains after disruptions"*, European Journal of Operational Research, Vol. 215, (2011).

[2] E. Albrechtsen and P. Hokstad. *"An analysis of barriers in train traffic using risk influencing factors"*, (in) Safety and Reliability, Swets & Zeitlinger, 2003, Lisse

[3] F. Auer and A. Schlöpp. *"Substanzermittlung der Oberbaukomponenten"*, ZEV Rail, 9/2012, (2012).

[4] J. Bates, J. Polak, P. Jones and A. Cook. *"The valuation of reliability for personal travel"*, Transportation Research Part E, Vol 37, (2001).

[5] C. R. Bhat and R. Sardesai. *"The impact of stop-making and travel time reliability on commute mode choice"*, Transportation Research Part B, Vol. 40, (2006).

[6] J. de Boer, B. van der Hoeven, M. Uittenbogaard, E. M. Dijkerman and W. Kruidhof. "Design based safety engineering applied to railway systems, part I", (in) Safety and Reliability, Swets & Zeitlinger, 2003, Lisse.

[7] J. de Boer, B. van der Hoeven, M. Uittenbogaard, E. M. Dijkerman and W. Kruidhof. "Design based safety engineering applied to railway systems, part II", (in) Safety and Reliability, Swets & Zeitlinger, 2003, Lisse.

[8] J. Braband. *"On the Justification of a Risk Matrix for Technical Systems in European Railways"*, FORMS/FORMAT, Part 3, (2011).

[9] R. Brkić and Z. Adamović. *"Research of Defects That Are Related with Reliability and Safety of Railway Transport System"*, Russian Journal of Nondestructive Testing, Volume 47/no. 6, (2011).

[10] G. Caimi, M. Fuchsberger, M. Laumanns and M. Lüthi: *"A model predictive control approach for discrete-time rescheduling in complex central railway station areas"*, Computers & Operations Research, Vol 39, (2012).

[11] V. Chandra, V. Kumar. *"Reliability and safety analysis of fault tolerant and fail safe node for use in a railway signalling system, Reliability Engineering and System Safety"*, Vol. 57, (1997).

[12] H.-K. Chen. "New models for measuring the reliability performance of train service", (in) Safety and Reliability, Swets & Zeitlinger, 2003, Lisse.

[13] R. M. Choudhry, D. Fang and S. Mohamed. "The nature of safety culture: A survey of the state-of-the-art", Safety Science, Vol. 45, (2007).

[14] A. Colini, P. Erto, M. Giorgio and A. Testa. *"A practical Markovian model of the availability and reliability of mass transport service with non-exponential repair times"*, (in) Reliability, risk and safety: Theory and Applications, Taylor & Francis, 2010, London.

[15] M. G. Cremonini, P. Lombardo, G. B. De Franchi, P. Paci, C. Rapicetta and L. Candeloro. *"Industrial areas and transportation networks risk assessment"*, (in) Safety and Reliability, Swets & Zeitlinger, 2003, Lisse.

[16] Y. Ding. *"Simulation model and algorithm for train speed regulation in disturbed operating condition"*, ZEV Rail, 10/2011, (2011).

[17] O. F. Dolven, B. H. Lindqvist, P. R. Hokstad. *"Statistical Modelling and Analysis of Failure and Inspection Data for a Railway Line"*, Proceedings of the European Safety and Reliability Conference, 2004, Berlin.

[18] R. Dorbritz. *"Methodology for assessing the structural and operational robustness of railway networks"*, PhD Thesis, 2012, Zurich.

[19] R. Dorbritz and U. Weidmann. *"Auswirkungen schwerer Störungen auf Bahnnetze"*, ZEV Rail, 6-7/2012, (2012).

[20]   J. Dyduch and M. Szczygielski. *"Model funkcjonalno-niezawodnościowy systemu SKZR"*, (in) Materiały Konferencji Zimowa Szkoła Niezawodności, 2008, Szczyrk.

[21]   D. Elms. *"Rail safety"*, Reliability Engineering & System Safety, Vol. 74, (2001).

[22]   S. Enjalbert, F. Vanderhaegen, M. Pichon, K. A. Ouedraogo and P. Millot. *"Assessment of Transportation System Resilience"*, Human Modelling in Assisted Transportation, (2011).

[23]   M. Enzi. *„Der optimale Re-Investitionszeitpunkt für das Gleis unter dem Aspekt der Lebenszykluskosten"*, ZEV Rail, 3/2012, (2012).

[24]   European Standard: EN50129

[25]   European Railway Agency. *"Intermediate report on the development of railway safety in the European Union"*, (2013).

[26]   L. H. J. Goossens, C. M. Pietersen and M. den Heijer-Aerts. *"Comparative quantitative risk assessment of railway safety devices"*, (in) Safety and Reliability, Swets & Zeitlinger, 2003, Lisse.

[27]   R. Goverde. *"Railway timetable stability analysis using max-plus system theory"*, Transportation Research Part B, Vol. 41, (2007).

[28]   R. Goverde. *"A delay propagation algorithm for large scale railway traffic networks"*, Transportation Research Part C, Vol. 18, (2010).

[29]   F. Hansemann and S. Marschnig. *"Der Gleisprophet – ein Impuls zur Nachhaltigkeit"*, ZEV Rail, 9/2012, (2012).

[30]   A. Hudoklin and V. Rozman. *"Reliability of railway traffic personnel"*, Reliability Engineering and System Safety, Volume 52, (1996).

[31]   A. Kadziński. *"Wprowadzenie do zagadnień bezpieczeństwa systemów kolejowych pojazdów szynowych"*, (in) Materiały XII Konferencji Naukowej Pojazdy Szynowe, 1996, Poznań-Rydzyna.

[32]   T. Krenželok, R. Briš, P. Klátil and V. Stýskala. *"Reliability and safety of railway signalling and interlocking devices. Reliability"*, (in) Risk and Safety: Theory and Applications, Tylor & Francis Group, 2010, London.

[33]   L. Kroon and D. Huisman. *"Algorithmic Support for Railway Disruption Management"*, Transitions Towards Sustainable Mobility, Part 3, (2011).

[34]   J. Kulczyk, T. Nowakowski and F. J. Restel. *"Application of phased-mission model to analyze reliability of combined rail-water transport system"*, Proceedings of the PSAM 11 & ESREL 2012 Conference, 2012, Helsinki.

[35]   B. Li. *"Markov Models for Bayesian Analysis about Transit Route Origin-Destination Matrices"*, Transportation Research Part B, Vol. 43, (2009).

[36]   C. Liebchen et al. *"Computing delay resistant railway timetables"*, Computers & Operations Research, Vol. 37, (2010).

[37]   B.-L. Lin, J.-W. Li and Y.-C. Huang. *"Train aggregation in a railway subsystem by Markov approach"*, International Journal of Modern Physics C, Vol. 19, (2008).

[38]   A. Lisnianski and I. Frenkel. *"Non-homogeneous Markov reward model for aging multi-state system under corrective maintenance"*, (in) Safety, Reliability and Risk Analysis: Theory, Methods and Applications, Tylor & Francis Group, 2005, London.

[39]   X. Liu, M. Saat and C. Barkan. *"Integrated risk reduction framework to improve railway hazardous materials transportation safety"*, Journal of Hazardous Materials, (2013).

[40]   R. van Loon, P. Rietveld and M. Brons. *"Travel time reliability impacts on railway passenger demand: a revealed preference analysis"*, Journal of Transport Geography, Vol. 19, (2011).

[41]   S. Marschnig and P. Veit. *"Life Cycle Management in der Realität"*, ZEV Rail, 9/2012, (2012).

[42]   W. Marsh and G. Bearfield. *"Using Bayesian Networks to Model Accident Causation in the UK Railway Industry"*, Proceedings of the European Safety and Reliability Conference, 2004, Berlin.

[43]   M. Młyńczak. *"Metodyka badań eksploatacyjnych obiektów mechanicznych"*, Oficyna Wydawnicza Politechniki Wrocławskiej, 2012, Wrocław.

[44]   T. Murray and T. H. Grubesic. *"Critical Infrastructure – Reliability and Vulnerability"*, Springer, 2007.

[45]   C. P. Neuman and N. M. Bonhomme. *"Evaluation of maintenance policies using Markov chains and fault tree analysis"*, IEEE Transactions on Reliability, 1975.

[46]   T. Nowakowski. *"Reliability Model of Combined Transportation System"*, Proceedings of the European Safety and Reliability Conference, 2004, Berlin.

[47] T. Nowakowski and M. Zając. *"Analysis of reliability model of combined transportation system"*, (in) Advances in Safety and Reliability, Tylor & Francis Group, 2005, London.

[48] B. Nyström and P. Söderholm. *"Improved railway punctuality by effective maintenance – a case study"*, (in) Advances in Safety and Reliability, Tylor & Francis Group, 2005, London.

[49] N. Olsson and H. Haugland. *"Influencing factors on train punctuality – results from some Norwegian studies"*, Transport Policy, Vol. 11, (2004).

[50] A. Pasquini, A. Rizzo and L. Save. *"Quantitative and qualitative analysis of SPAD"*, Proceedings of the European Safety and Reliability Conference, 2002, Lyon.

[51] L. Podofillini, E. Zio, M. Marella. *"A multi-state Monte Carlo simulation model of a railway network system"*, (in) Advances in Safety and Reliability, Tylor & Francis Group, 2005, London.

[52] G. Potthoff. *"Verkehrsströmungslehre (Band 3) – Die Verkehrsströme im Netz"*, Transpress, 1964, Berlin.

[53] H. Radbo, B. Renck and R. Andersson. *"Feasibility of railway suicide prevention strategies: A focus group study"*, (in) Advances in safety, reliability and risk management, CRC Press/Balkema, 2012.

[54] H. Radbo, I. Svedung and R. Andersson. *"Suicide and the potential for suicide prevention on the Swedish rail network: A qualitative multiple case study"*, (in) Advances in safety, reliability and risk management, CRC Press/Balkema, 2012.

[55] F. Renpenning, J. Braband and S. Wery. *"Application of functional hazard assessment in railway signalling"*, (in) Safety and Reliability, Swets & Zeitlinger, 2003, Lisse.

[56] F. Renpenning. *"Reliability Prediction in Railway Signalling"*, Proceedings of the European Safety and Reliability Conference, 2004, Berlin.

[57] F. J. Restel. *"Measures of reliability and safety of rail transportation system"*, (in) Advances in safety, reliability and risk management, CRC Press/Balkema, 2012.

[58] P. Rietveld, F. R. Bruinsma and D. J. van Vuuren. *"Coping with unreliability in public transport chains: A case study for Netherlands"*, Transportation Research Part A, Vol. 35, (2001).

[59] M. Saat and C. Barkan. *"Generalized railway tank car safety design optimization for hazardous materials transport"*, Journal of Hazardous Materials, Volume 189, (2011).

[60] A. Schöbel and T. Maly. *"Operational fault states in railways"*, European Transportation Research Review, Springer, 2012.

[61] S. Schwartz. *"Identifikation von Sicherheitsbarrieren am Bahnübergang"*, ZEV Rail, 1-2/2010, (2010).

[62] J. Törnquist and J. A. Persson. *"N-tracked railway traffic re-scheduling during disturbances"*, Transportation Research Part B, Vol. 41, (2007).

[63] J. Törnquist-Krasemann. *"Design of an effective algorithm for fast response to the re-scheduling of railway traffic during disturbances"*, Transportation Research Part C, Vol. 20, (2012).

[64] H. Ugajin. *"Human Factors Approach to Railway Safety"*, QR (Quarterly Report) of RTRI (Railway Technical Research Institute), Vol. 40, (1999).

[65] P. Vansteenwegen and D. Van Oudheusden. *"Decreasing the passenger waiting time for an intercity rail network"*, Transportation Research Part B, Vol. 41, (2007).

[66] M. Vromans. *"Reliability of Railway Systems"*, TRAIL Thesis series T2005/7, 2005.

[67] M. Vromans, R. Dekker and L. Kroon. *"Reliability and heterogeneity of railway services"*, European Journal of Operational Research, Vol. 172, (2006).

[68] C. Wietfeld and C. H. Rokitansky. *"Markov chain analysis of alternative medium access control protocols for vehicle roadside communications"*, Proceedings of Vehicular Technology Conference, 1995.

[69] J. R. Wilson and B. J. Norris. *"Human factors in support of a successful railway: a review"*, Cognition, Technology & Work, Volume 8, (2005).

[70] W. Zamojski. *"Markowski model niezawodności systemu transportu dyskretnego"*, Materiały Konferencji Zimowa Szkoła Niezawodności, 2001, Szczyrk.

[71] W. Zamojski et al. *"Systemy transportu dyskretnego – modele, niezawodność"*, Wydawnictwa Komunikacji i Łączności, 2007, Warszawa.

[72] Y. Zheng et al. *"Carrying Capacity Reliability of Railway Networks"*, Journal of Transportation Systems Engineering and Information Technology, Vol. 11, (2011).

Analysis of interdependencies of the Mexico City Metro System

Jaime Santos-Reyes[a*], and Diego Padilla-Pérez[a]

[a] SARACS Research Group, SEPI-ESIME, IPN, Mexico City, Mexico

Abstract: The Mexico City Metro underground system has been regarded as the second largest Metro in North America after the New York City Metro. It is believed that in 2006 the system served over one billion passengers, the fifth highest in the world. Given this, a threat to the Metro transport system may either have an impact on other industries that rely on it or to the other modes of transportation in the City. Interdependencies amongst the key components of the Metro system, therefore, must be understood and adequately addressed. The paper addresses the modelling of the interdependencies amongst the Metro lines by applying a 'Systemic Safety Management System' (SSMS) model. The paper gives an account of the ongoing research project.

Keywords: Interdependency, Metro System, Mexico City, SSMS Model.

1. INTRODUCTION

1.1. Critical infrastructures & Its Context

The concepts and the importance of "interdependencies" and "critical infrastructures" took interest after the publication of the report "Critical Foundations: Protecting America's Infrastructures", the report of the U.S. President's Commission on Critical Infrastructure Protection (PCCIP) [1]. The PCCIP has defined an "infrastructure" as a network of independent, mostly privately-owned, man-made systems and processes that function collaboratively and synergistically to produce and distribute a continuous flow of essential goods and services [1]. Moreover, the Commission focused on eight critical infrastructures; i.e., telecommunications, electric power systems, natural gas and oil, banking and finance, transportation, water supply systems, government services, and emergency services. More recently, the Critical Infrastructure Assurance Office (CIAO), an interagency office created under Presidential Decision Directive (PDD) 63 to assist in coordinating the federal government's initiatives on critical infrastructure protection. The CIAO defined infrastructure as "the framework of interdependent networks and systems comprising identifiable industries, institutions (including people and procedures), and distribution capabilities that provide a reliable flow of products and services essential to the defence and economic security of the United States, the smooth functioning of governments at all levels, and society as a whole". Moreover, the CIAO's report included the following critical infrastructures: food/agriculture (production, storage, and distribution), space, numerous commodities (iron and steel, aluminium, finished goods, etc.), the health care industry, and the educational system [2].

A number of studies have been conducted on interdependent critical infrastructures. For example, the International Risk Governance Council (IRGC) has conducted research on both the risks associated with five individual infrastructures and the risk associated with the increasing interdependency between them. The infrastructures considered by the IRGC were {1} electric power and gas supply; {2} information and communication services -as provided by the internet as well as ICT to monitor and control other infrastructures; {3}urban water supply and waste water treatment; {4} rail transport. [3,4].

Rinaldi et al. [5] has proposed four classes of interdependencies; i.e.: 'physical', 'cyber', 'geographic', and 'logic'. 'Physical' interdependency occurs when two infrastructures are physical interdependent if the state of each is dependent on the material output(s) of the other. An infrastructure has a 'cyber'-

---

* E-mail address: jrsantosr@hotmail.com

interdependency if its state depends on information transmitted through the information transmitted. 'Geographic' interdependency, on the other hand, occurs when infrastructures are geographically interdependent; i.e., if a local environmental event can create state changes in all of them. Finally, 'logical' interdependencies occurs when infrastructures are logically interdependent if the state of each depends on the state of the other via a mechanism that is not physical, cyber, or geographical connection. A number of research has been conducted on the survivability and vulnerability of infrastructure systems [6-17].

1.2. Examples of Critical Infrastructures

Any disruption or destruction of key technical systems could have significant consequences; for example, significant impact on public health and safety, public confidence, negative impacts to the environment, the economy. Many of these systems house significant amounts of hazardous materials, fuels, and chemical catalysts that enable important production and processing functions. A brief description of some of the following systems is presented in the subsequent section: water, telecommunications, energy, and transport.

1.2.1 Energy

Energy may be regarded as one of the key infrastructures in the modern society. The energy sector is commonly divided into two segments in the context of critical infrastructure protection: electricity and oil and natural gas [5]. It is believed the electric industry services almost 130 million households and institutions. The United States, for example, consumed nearly 3.6 trillion kilowatt hours in 2001. Moreover, every form of productive activity (e.g. businesses, manufacturing plants, schools, hospitals, or homes) requires electricity. [1,5].

1.2.2 Water

Any Nation's water sector is critical from both a public health and an economic standpoint. The water sector consists of two basic, yet vital, components: fresh water supply and wastewater collection and treatment. Sector infrastructures are diverse, complex, and distributed, ranging from systems that serve a few customers to those that serve millions. These utilities depend on reservoirs, dams, wells, and aquifers, as well as treatment facilities, pumping stations, aqueducts, and transmission pipelines. [1,5].

1.2.3 Telecommunications

The telecommunications sector provides voice and data service to public and private users through a complex and diverse public-network infrastructure encompassing the Public Switched Telecommunications Network (PSTN), the Internet, and private enterprise networks. The PSTN provides switched circuits for telephone, data, and leased point-to-point services. Because the government and critical infrastructure industries rely heavily on the public telecommunications infrastructure for vital communications services, the sector's protection initiatives are particularly important. [1,5].

1.2.4 Transport

The transportation sector consists of several key modes: aviation, maritime, rail, road and public mass transit. As a whole, the various transportation modes provide mobility of the population and contribute to the quality of life of any country's inhabitants. Given the above, a threat to the transportation sector may either have an impact on other industries that rely on it or to the other modes of transportation. Interdependencies amongst modes of transportation, therefore, must be adequately addressed. [1,5].

2. THE MEXICO CITY METRO SYSTEM

The Mexico City Metro underground system has been regarded as the second largest Metro in North America after the New York City Metro. It is believed that in 2006 the system served over one billion passengers, the fifth highest in the world. The Metro system map showing the eleven lines is shown in Fig1; the Metro system comprises a total of 175 stations and 106 underground stations. Every Metro line transports hundreds of thousands of users every single day. [18].

Figure 1: The Mexico City Metro system. [18].

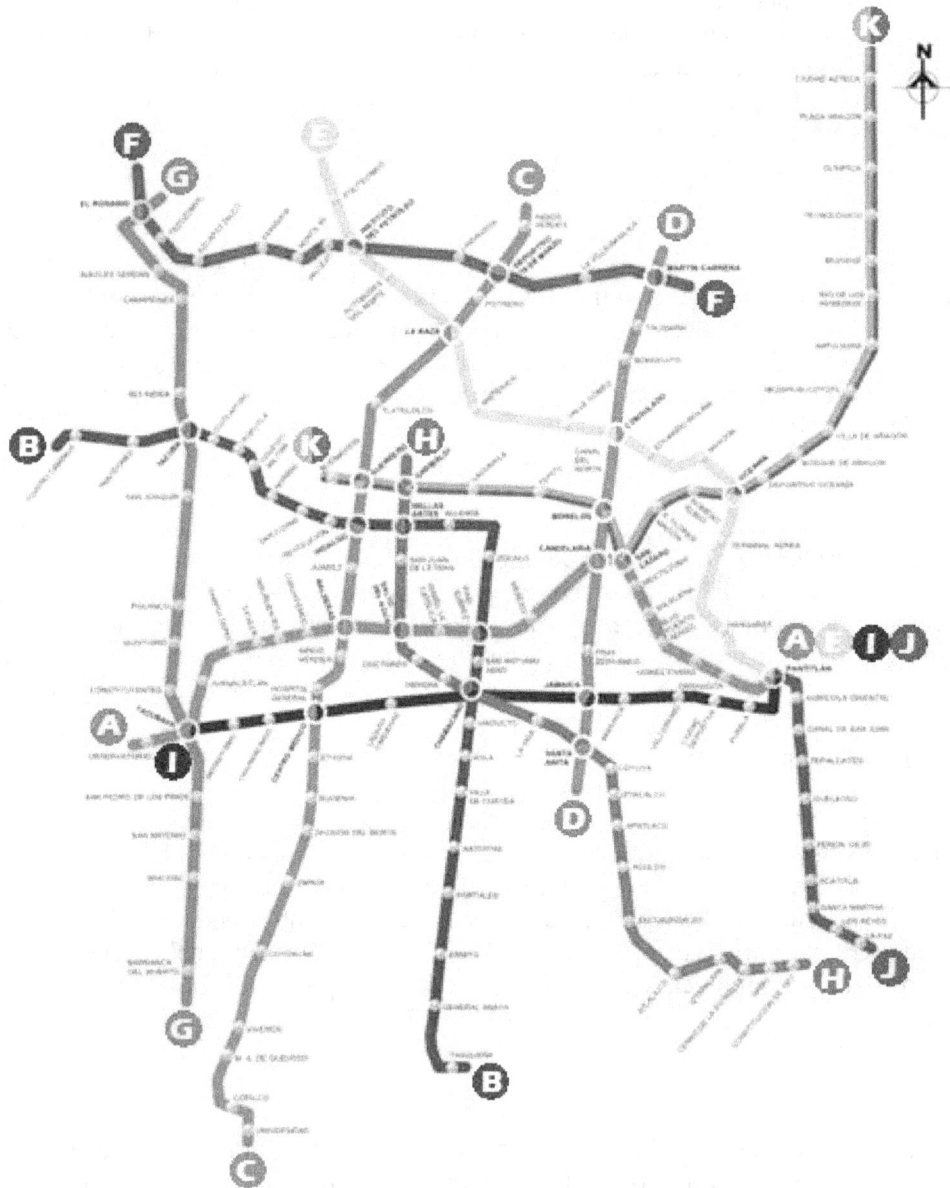

As mentioned in the introduction section, any disruption in any component of a highly interrelated system may cause a 'domino' or cascading effect. For example, on 22nd January 2008, a failure in the power supply system to the Metro system caused a disruption and affected several lines within the system; the incident caused trains to halt in 68 stations and it is believed the incident affected about

83,000 users. Also, the public transport could not cope with the amount of users being affected by the disruption. [19].

3. A SSMS MODEL

A 'SSMS' model is intended to maintain risk within an acceptable range in any organization's operations. The model is proposed as a structure for an effective safety management system. It may be argued that if all the sub-systems and connections are present and working effectively, the probability of a failure should be less than otherwise. A brief description of the 'structural organization' of the model is given as follows: this systemic approach to safety management consists of a set of five necessary and sufficient interrelated subsystems, labelled as systems 1 to 5. *System 1, safety policy implementation*, consists of various operations of an organization in which the organization's safety policy must be implemented. *System 2, safety co-ordination*, ensures that the various operations of system 1 operate in agreement. *System 3, safety functional*, ensures that system 1 implements the organization's safety policies. *System 3*, safety audit*, is part of system 3 and it is concerned with safety sporadic audit. System 4, safety development, is responsible for identifying strengths, weaknesses, threats, and opportunities that can suggest systemic changes to the organization's safety policies. *System 4*, confidential report*, is part of system 4 and it is concerned with confidential reports or causes of concern that may require direct and immediate intervention of the corporate management. Finally, *system 5, safety policy*, is responsible for establishing safety policies for the whole organization. A full description of the model is given elsewhere [20-22].

Figure 1: Recursive structure of a 'SSMS' model. (© Santos-Reyes).

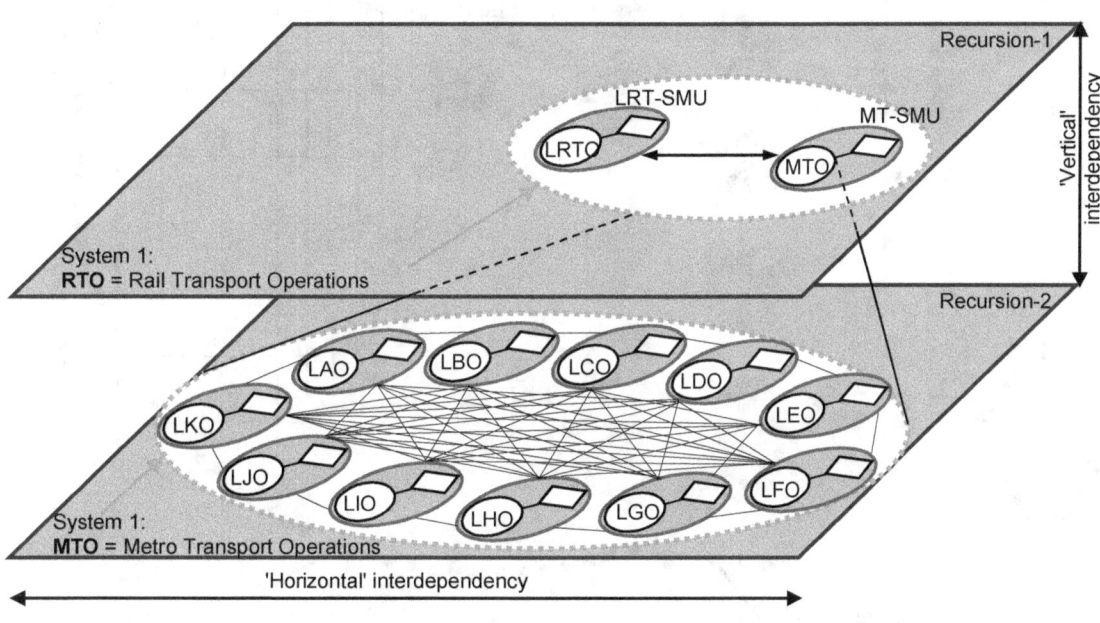

Acronyms:

MT-SMU=Metro Transport-Safety Management Unit;
MTO= Metro Transport Operations
L$_{A-K}$-SMU= Line A-K Safety Management Unit;
L$_{A-K}$O= Line A-K Operations

Figure 3: 'Horizontal' & 'vertical' interdependency. (© Santos-Reyes).

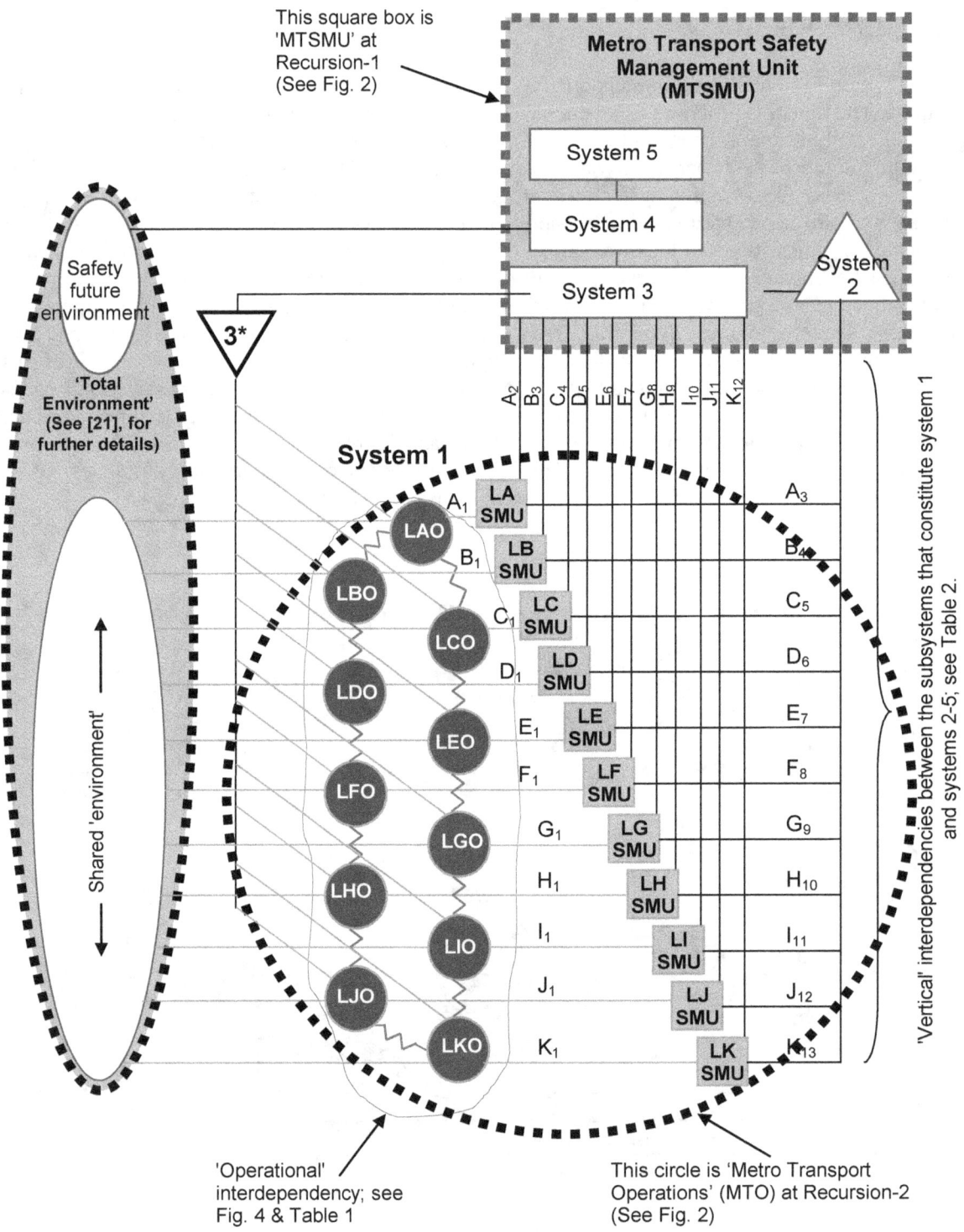

Note: see Fig. 2 for details of the acronyms used in the figure.

Figure 4: 'Operational' interdependency. (© Santos-Reyes).

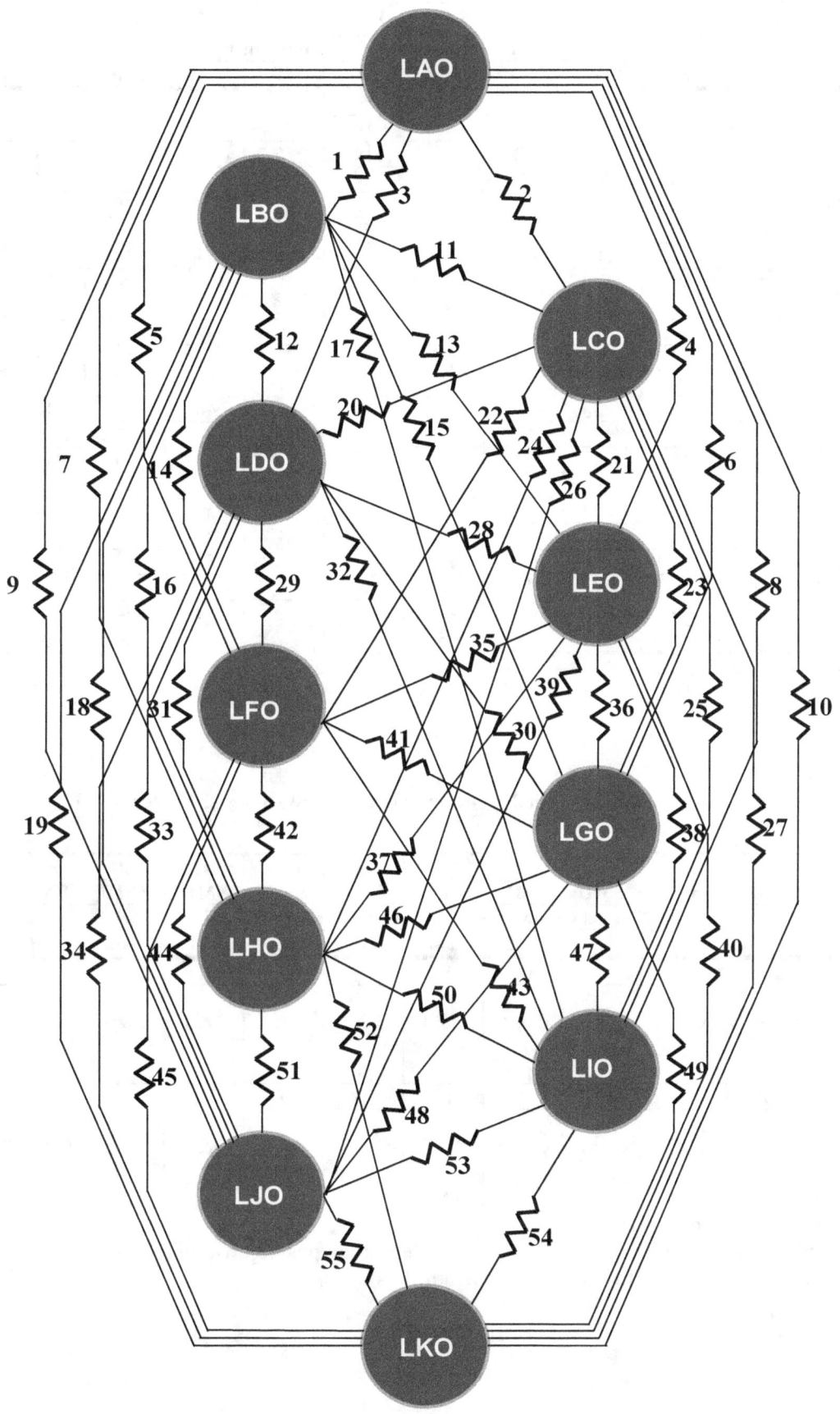

‿‿‿ represents the interdependencies which may be 'strong' or 'weak'; see Table 1.

Table 1: 'Operational' interdependency amongst the Metro Lines.

	LAO	LBO	LCO	LDO	LEO	LFO	LGO	LHO	LIO	LJO	LKO
	Number of Zigzag lines shown in Fig. 4										
	S/W	S/W	S/W	S/W	S/W	S/W	S/W	S/W	S/W	S/W	S/W
LAO		1	2	3	4	5	6	7	8	9	10
		S	S	W	W	S	S	W	S	S	S
LBO	1		11	12	13	14	15	16	17	18	19
	S		S	W	W	W	S	S	S	W	W
LCO	2	11		20	21	22	23	24	25	26	27
	S	S		W	S	S	W	W	S	W	S
LDO	3	12	20		28	29	30	31	32	33	34
	S	W	W		S	S	W	S	S	W	S
LEO	4	13	21	28		35	36	37	38	39	40
	S	S	S	S		S	W	W	S	S	S
LFO	5	14	22	29	35		41	42	43	44	45
	W	W	S	S	S		S	W	W	W	W
LGO	6	15	23	30	36	41		46	47	48	49
	S	S	W	W	W	S		W	S	W	W
LHO	7	16	24	31	37	42	46		50	51	52
	S	S	W	S	W	W	W		S	W	S
LIO	8	17	25	32	38	43	47	50		53	54
	S	S	S	S	S	W	S	S		S	W
LJO	9	18	26	33	39	44	48	51	53		55
	W	W	W	W	W	W	W	W	W		W
LKO	10	19	27	34	40	45	49	52	54	55	
	S	S	S	S	S	W	W	S	W	W	

S='Strong' interdependency.
W='Weak' interdependency.

Table 2: 'Vertical' interdependency between System1 & Systems 2-3.

	LA-SMU	LB-SMU	LC-SMU	LD-SMU	LE-SMU	LF-SMU	LG-SMU	LH-SMU	LI-SMU	LJ-SMU	LK-SMU
	No. Channel Communication & Control connecting Systems 2&3 (Fig. 3)										
	ToI	ToI	ToI	ToI	ToI	ToI	ToI	ToI	ToI	ToI	ToI
System 2	A_3	B_4	C_5	D_6	E_7	F_8	G_9	H_{10}	I_{11}	J_{12}	K_{13}
	C & L	C & L	C & L	C & L	C & L	C & L	C & L	C & L	C & L	C & L	C & L
System 3	A_2	B_3	C_4	D_5	E_6	F_7	G_8	H_9	I_{10}	J_{11}	K_{12}
	C & L	C & L	C & L	C & L	C & L	C & L	C & L	C & L	C & L	C & L	C & L

ToI=Type of Interdependency.
C= 'Cyber' (e.g. shared information related to the Metro train position within the Line).
L= 'Logic' (e.g. decisions being taken yb the controllers).

4. MODELLLING INTERDEPENDECY FOR THE CASE STUDY

4.1. The Modelling Process

In order to model interdependencies, an 'infrastructure' has been considered as a 'system'; i.e., any entity which consists of interdependent 'parts'. Given this, two levels of recursion for the present case study are shown in Fig. 2. It can be seen that System 1 at level 1 contains the sub-system of interest; *i.e.,* 'Metro Transport Operations' and its associated Safety Management Unit (SMU); i.e., 'Metro Transport Safety Management Unit' ('MT-SMU').

Increasing the level of resolution of the system of interest, *i.e.,* 'MTO' at one level below recursion 1 will result in a system that contains the eleven lines that constitute the whole Metro system (i.e., system 1), at the level of recursion 2. The 'systems' identified in Fig. 2 have been represented in the format of the 'structural organization' of the model; i.e. systems 1-5 and their associated connections as shown in Fig. 3.

The following types of interdependencies have been identified in Fig. 3: {a} 'operational' ('circles'), see Table 1; {b} 'managerial' ('square boxes'); and {c} 'environmental' ('elliptical' symbol; these are not discussed here). These interdependencies occur 'horizontally' at one level of recursion only; i.e. at recursion 2.

'Vertical' interdependencies have been identified that they occur only between levels of recursions; in this particular case, between Systems 2-5 and system 1 (i.e. the eleven lines of the Metro system), as shown in Figs. 3, 4 & Table 2.

4.2. Summary of the Type of Interdependency Highlighted by the Model

4.2.1 'Vertical' interdependency

The principle of *recursion* has been proved to be a powerful concept in identifying 'vertical' and 'horizontal' interdependency. ('Recursion' may be understood as a 'system' contains and it is contained by another 'system'). 'Vertical' interdependency occur between two levels of recursions; 'horizontal' interdependencies, on the other hand, occur at every level of recursion (see Figs. 2-4). The identified systems were mapped onto the format of the 'structural organization' (i.e. systems 1-5) of the model (Fig. 3). The connections between every 'SMU' ('square boxes') with Systems 2&3 indicate the 'vertical' interdependencies; the nature of the relationship is managerial; see Table 2.

4.2.2 'Horizontal' interdependency

'Horizontal' interdependency amongst the collection of subsystems that constitute System 1 occur at every level of recursion. Three types of interdependencies have been identified: 'operational', 'managerial', and 'environmental'.

'Operational' interdependencies

'Operational' interdependency (circles) could be 'strong' or 'weak' (Fig. 4). A 'strong' interdependency implies that the operations are highly interdependent; for example, the 'correspondence' stations between two Metro lines has been considered here as a 'strong' interdependency. A 'weak' interdependency, implies a relatively 'weak' dependency; for example, the Metro lines with no 'correspondence' between lines were considered within this category. The results are shown in Fig. 3 & Table 1.

'Managerial' interdependency

The safety management units ('the square boxes') stands for the operations ('circle') that the management unit is supposed to regulate: the operations are 'under its control' (see the connections between every SMU and the operation shown in Fig. 3). However, when a disturbance occur in any of the SMU's operations, this in turn will propagate to the whole system. In order to bring the disturbance under control the communication amongst the SMUs becomes essential to achieve this. This is not shown here.

'Environmental' interdependency

The 'environment' is multi-dimensional; i.e. economical drivers; it contains of a market, of a supply-situation, of customers of users, of the general public, natural hazards. (See Figures 4&7). In general, the 'environmental' factors are those that threaten the 'system'; for example, the cascading effects caused by the earthquake and tsunami of March 11, 2011 in Japan. [23]. This is not shown here.

5. CONCLUSION

The paper has presented a 'systemic' approach to model interdependencies for the case of Mexico City Metro system; i.e., by applying a 'SSMS' model. It has been found that interdependencies occur 'vertically and 'horizontally'. 'Vertical' interdependencies occur between two levels of recursions. 'Horizontal' interdependencies occur at every level of recursion; i.e.: {a} 'operational'; {b} 'managerial; and {c}'environmental'.

More research is being conducted on: a) identifying interdependency amongst the SMUs (Fig. 3); b) identifying the 'environmental' interdependency (Fig. 3); c) Identifying the interdependency between 'operations' (circle) and their associated 'management units' (square boxes) as shown in Fig. 3; and d) modelling the 'emergency response' in the context of the 'SSMS' model in case of an emergency situation such as the example described briefly in section 2.

Acknowledgements

The research project was funded by SIP-IPN-No. 20141500.

References

[1]   PCCIP (President's Commission on Critical Infrastructure Protection), *"Critical Foundations: Protecting America's Infrastructures."* [Online]. Available: http://www.ciao.gov., (2007), USA.

[2]   Presidential Decision Directive 63 (PDD). [Online]. Available: http://www.ciao.gov. (2009), USA.

[3]   IRGC, *"White paper on managing and reducing social vulnerabilities from coupled critical infrastructures"*, (2006), Geneva.

[4]   IRGC, *"Policy brief on Managing and reducing social vulnerabilities from coupled critical infrastructures"*, (2007), Geneva.

[5]   S.M. Rinaldi, J. P., Peerenboom, and T.K., Kelly. *"Identifying, Understanding, and Analyzing Critical Infrastructure Interdependencies."* IEEE Control Systems Magazine, vol. 21, pp. 11-25. (2001).

[6]   A. Balakrishnan, T.L., Magnanti and P. Mirchandani. *"Designing Hierarchical Survivable Networks."* Operations Research, vol. 46, pp. 116-136, January- February. (1998).

[7]   S. Chamberland and B. Sanso. *"On the Design of Multitechnology Networks,"* INFORMS Journal on Computing, vol. 13, pp. 245-256. (2001).

[8] Y.Y. Haimes, N.C. Matalas, J.H. Lambert, B.A. Jackson and J.F.R. Fellows. *"Reducing Vulnerability of Water Supply Systems to Attack,"* Journal of Infrastructure Systems, vol. 4, pp. 164-177, December. (1998).

[9] T.A. Longstaff and Y.Y. Haimes. *"A Holistic Roadmap for Survivable Infrastructure Systems"*, IEEE Transactions on Systems, Man and Cybernetics, vol. 32, pp. 260-268, March. (2002).

[10] Y.Y. Haimes and P. Jiang. *"Lontief-based Model of Risk in Complex Interconnected Infrastructures"*, Journal of Infrastructure Systems, vol. 7, pp. 1-12, March. (2001).

[11] M. Amin. *"Toward Self-healing Energy Infrastructure Systems"*, IEEE Computer Applications in Power, vol. 14, pp. 20-28, January. (2001).

[12] E.E. Lee, D. Mendonca, J.E. Mitchell and W.A. Wallace. *"Restoration of Services in Interdependent Infrastructure Systems: A Network Flows Approach,"* Technical Report 38-03-507, Decision Sciences and Engineering Systems, Rensselaer Polytechnic Institute, Troy, NY, (2003), US.

[13] V. Cozzani, G. Gubinelli, G. Antonioni, G. Spadoni and S. Zanelli. *"The assessment of risk caused by domino effect in quantitative area risk analysis"*. Journal of Hazardous Materials;127:14–30. (2005).

[14] E. Cagno, M. De Ambroggi, O. Grande and P. Trucco. *"Risk analysis of underground infrastructures in urban areas: time-dependent interoperability analysis"*. Reliability, risk and safety: theory and applications, In: Bris, Guedes Soares, Martorell, editors. ESREL Conference, (2009).

[15] E. Cagno, O. Grande and P. Trucco. *"Towards an integrated vulnerability and resilience analysis for underground infrastructures."* In: Proceedings of the third resilience engineering symposium, Antibes—France, (2008).

[16] W. Kroger. *"Critical infrastructures at risk: a need for a new conceptual approach and extended analytical tools"*. Reliability Engineering 6 System Safety, 93, pp. 1781-1787. (2008).

[17] T. Aven. *"Identification of safety and security critical systems and activities"*. Reliability Engineering and System Safety, Vol. 94, issue 1, pp. 404-411. (2008).

[18] The Mexico City Metro. Accessed at: www.metro.df.gob.mx/. (2014). Mexico.

[19] D. Padilla-Pérez. *"Modelado de interdependencias de sistemas críticos: Caso del Sistema de transporte Metro de la Cd. de México"*. Tesis Doctoral. Departamento de Ingeniería de Sistemas, SEPI-ESIME, IPN, (2014). México.

[20] J. Santos-Reyes and A.N. Beard. *"A systemic approach to fire safety managing. Fire Safety Journal"*. Vol. 36, pp. 359-390. (2001).

[21] J. Santos-Reyes and A.N. Beard. *"A systemic approach to managing safety"*. Journal of Loss Prevention in the Process Industries. Vol. 21, No. 1, pp. 15-28. (2008).

[22] J. Santos-Reyes and A.N. Beard. *"A systemic analysis of the Edge-Hill railway accident"*. Accident Analysis and Prevention, vol. 41, No. 6, pp. 1133-1144. (2009).

[23] NJD (The National Diet of Japan*). "The Fukushima nuclear accident independent investigation commission"*. 2012. Japan.

BOTTLENECKS OF INLAND CONTAINER TERMINALS

Mateusz Zajac, Franciszek J. Restel[a],
[a] Wroclaw University of Technology

Abstract: Availability of the intermodal transport chain depends on the proper functioning of the container terminals, including their ability to perform cargo handling infrastructure, cost- effectiveness and scope of services, quality and reliability. Increasing number of intermodal operators make that competitiveness becomes crucial issue to survive in the market. New objectives and performance measures need to be identified and employed to evaluate the performance of a container terminal. The aim of article is to show the most important elements of container warehousing and its impact on process availability and productivity. The article threats problems on inland intermodal terminals.

Proper operation of the intermodal transport chain depends on the proper functioning of the terminals, including their ability to perform cargo handling infrastructure, cost- effectiveness and scope of services, quality and reliability. Modern combined transport terminal is more than a simple transshipment point. It develops in the creation of centers freight with a wide range of services.

Simulation tests confirmed that the storage process guided by specific rules may result in a significant reduction in energy demand in the course of handling containers. Depending on the size of the node handling, the degree of use and speed of movement of the intermodal units can be a saving of up to 50%.

This article is a presentation of the progress of work on the project , which aims to develop a practical method that allows to use the knowledge to creating the functionality of intermodal transport terminal, taking into account the characteristics of its work, including efficiency, effectiveness, reliability, safety, ecology.

Keywords: Container terminal availability, terminal operation, terminal maintenance

1.   MOTIVATION

After the difficult years of 2008-2010 container transport becomes stronger. In Poland it is visible seeing more and more new container handling transshipment points. Currently, the total container turnover in the market of intermodal transport is estimated by the owners of intermodal companies over 2.0 million TEUs per year (for comparison, in 2007, there are approximately 1.4 million TEU).

Facing of the White Paper on Transport recommendations (document leading transport policy in Europe) the increase of intermodal transport seems to be natural phenomena, nevertheless as a logistics process requires scientific work and research. The effect of which is to strengthen the competitiveness of intermodal transport to the traditional road to the carriage of highly processed goods.

Modern combined transport terminal is more than a simple transshipment point. It develops in the creation of centers freight with a wide range of services [3], [5]. Sea terminals are to be run in such a way as to reduce to a minimum residence time of the loading units within the terminal. Using sophisticated technology, handling, such as full automation of the process can substantially reduce the time of cargo handling, eliminate errors, increase the level of safety and reliability[1], [2], [4]. These technologies are extremely expensive and are not widely used. In smaller ports, high throughput is achieved by streamlining operations. The inland terminals link transport and storage functions. Problem in intermodal transshipment hubs is linked with choosing appropriate method of container warehousing [10], [11]. Very often it is necessary to move container several times from one point to another during process of storage. The results are more expensive container service and probability that containers can't be easy available then are needed.

The increase in intermodal transport is a natural phenomenon in the face of the *White Paper on Transport* recommendations. Nevertheless, intermodal transport as a logistics process requires scientific work and research, the effect of which is to strengthen the competitiveness of intermodal transport to the traditional road to the carriage of highly processed goods.

Intermodal technology functioning is mainly based on experience in Poland. Over the years, intermodal transport was underrated way of transporting goods. Its principal advantage lies in combining functionality with the ability to cargo transport and storing in intermodal transshipment point. Intermodal transport technologies are shown in [6]. Design rules container terminals are presented in the [9]. Both publications are land-based container terminal.

Both the technology and design rules terminals do not show know how to manage the movement of cargo units inside the terminal. However, the functioning of the inland container terminals is far different from the typical container ports on which there is a lot of information in foreign literature [12].

Sea terminals are to be run in such a way as to reduce to a minimum residence time of the loading units within the terminal [8]. This is due to the need for high bandwidth as a result of the conditions established infrastructure and container turnover. Using sophisticated technology, handling, such as full automation of the process can substantially reduce the time cargo handling, eliminate errors, increase the level of safety of the process. These technologies are extremely expensive and are not widely used. In smaller ports, high throughput is achieved by streamlining operations. One of the solution is increasing tariff for storage of cargo at the port [15].

The inland terminals, as mentioned previously, links Transport and storage functions. In this case, the tariff for the storage of empty containers or loaded is decreasing. Both types of containers are stored within a storage space. Problem in intermodal transhipment hubs to adopt an appropriate method of storage of intermodal units, the implementation process container depots, so that there was no need of their translocation to another storage location. In reality Polish intermodal hubs, stacking containers and large volume of financial and intuitive decision-making, such situations often occur. This is the reason for the formation of additional costs and sometimes even necessary, adjusting the container several times.

In the international literature, little space is devoted to the theme of inland terminals. Generally it is a showcase of new technologies intermodal attempt to analyze their applicability, detailed technical solutions. There is no literature on the process of storage. There have been no analysis of the arguments has to be taken into account when storing. Do not analyzed the information contained in the transport documents for their use in the management of places components in intermodal transshipment node. We can say that this area of knowledge is not recognized, and the practice sets the rules in force here.

The remainder of this article outlines basic procedures performed during the reception and dispatch of goods to and from the container terminal. The principles of selecting the places of storage containers, and indicated the formation of any disruption to the service container terminal.

2.   THE NEED FOR QUALITY IN USE OF INTERMODAL TRANSPORT

The proper functioning of the whole chain of intermodal transport depends largely on the proper functioning of the terminal, including first and foremost on their ability to perform cargo handling infrastructure, cost- effectiveness and scope of services, quality and reliability. Modern combined transport terminal is more than a simple transshipment point. It develops in the creation of centers freight with a wide range of services.

EU report [13] the most characteristic tendencies in the development of combined transport terminals in Europe include, among others:
• transition from isolated terminals to integrated logistic centers freight handling the " hub " of the terminal for combined transport as a key location in the center, while the development of a network of medium and even small terminals with maximum loading process automation;
• concentration of resources to improve service processes in terminals, especially automated cargo handling, cargo handling technology standardization and implementation of complex information technology systems .

Combined Transport Development Strategy in Poland (Published in 2004) shows such the need to develop innovative technologies for intermodal terminals. In turn, released in 2006 by the General Department for Energy and Transport of the European Commission's publication " In search of the slide Intermodality efficiency growth" simply points to the need for research projects focused on solving the problems of intermodal transport in order to increase its efficiency and wider dissemination.

Mentioned basic needs include:

• improving the quality of services with a focus on the creation of information technologies in the management of terminals,

• looking for ways to achieve synergies between the supply chain for intermodal transport

• further harmonization and interoperability between transport modes, manifested in new technologies, transport and handling of intermodal loading units.

Among the solutions sought European Commission proposes to focus on the work of, among others:

• identify problems and bottlenecks in the operation of intermodal terminals

• identification of tasks to increase efficiency of intermodal terminals , and these tasks with the help of state institutions and businesses.

Conducted by the author identification of multimodal transport operators in Poland have indicated that postulates presented above are still very relevant. For many years, one can observe the dynamic growth of transported cargo units in Poland related to intermodal transport . They are expected to be a solution that will allow for further sustainable development of transport in Poland. These are primarily solutions for qualitative and quantitative evaluation of the work, analytical tools, and tools that enable better management of the company in the operational field.

3.   CHARACTERISTICS OF PROCEDURES FOR ENTRY AND CONTAINER

Most intermodal trans-shipment hub located inland supports two modes of transport : road and rail transport (in CEE is difficult to discern the terminal supports three modes of transport) . Thus, in the following section describes the procedures for entry and exit of containerized cargo by rail and road.

The procedure for the adoption of the terminal

Container train is the most Anticipated few days prior to arrival at the terminal. Notification is sent by e-mail in the form of a list containing the most common train number, wagon number, the number of the container located on the wagon, the gross weight of the container, information on the place of delivery, round trip, Incoterms rule, etc.

Composition station when you arrive at the container terminal is checked for possible damage units (document interchange) . If the damage is not aware of the physical follows the adoption of containers on the terminal - and start unloading of storage.

Similarly, the procedure will adopt the container by road. Adoption of the container terminal road transport is also most frequently preceded by notification . This time, most workers make advising terminal based on the order issued by a company that provides container terminal . After the formal notification of the completion of the notification and the safety formulas tractor -trailer carrying a container can enter the loading area. The driver shall be submitted together with the accompanying documents to the person carrying out the inspection of container for damage unit load. Then, if no evidence of deficiencies is awarded to the terminal. The driver is required to set up a set in the space where the container is picked up.

Upon notification by the trustee cargo container needs material is taken from the site and loaded onto a semi-trailer containers. Then , before leaving the terminal , the driver is equipped with special set of transport documents. The driver also receives an instruction from the load unit and therefore an indication of the customs , trance journey to the customer and the procedures for the transfer of cargo transport. The driver is well informed about the details of the date for delivery of the cargo.

4.   PRINCIPLES IN CONTAINER STORAGING

The ground terminals loaded containers are stored mostly to three layers to four layers vain. The storage yard in a terminal is usually divided into rectangular regions called storage blocks or blocks. A typical block has seven rows (or lanes) of spaces , six of Which are used for storing containers in stacks or columns, and the seventh reserved for truck passing . Each row typically Consists of over twenty 20 -ft container stacks stored lengthwise end to end . For storing a 40 -ft container stack , two 20 -ft stack spaces are used.

Load distribution and hence the allocation of storage is done by machine operators . This is done on the basis of their experience and relying on the information derived from public goods . Basic information to be taken when allocating loads are:

• whether the container is empty, loaded , refrigerated, tank , ADR / RID;
• the size of the container,
• the expected storage time charge on the terminal ,
• the recipient,
• the gross mass,
• operator.

The essential art of machine operators to memorize and consistently putting in a storage container so as not to turn the download does not require adjustment of the upper layers of containers . This task is difficult and involves unreliability . This problem increases the lack of information from one of the main customers of schedule downloads terminal at the time of arrival of containers by rail to unload the cargo units . As a result, the containers are unloaded in free space components (cached) , and after being informed of the date of delivery segregated and placed in the correct order.

Schedules cargo operations are difficult to define because they vary depending on the sender or recipient, traffic conditions , etc. , although this time the service should be as short as possible. Therefore, previous planning manual handling is often very difficult , if not impossible , due to random factors beyond the control of the operator terminal. To transfer the containers in the number of hours between scheduled transport services large batches , the temporary storage of containers is essential. Meets the buffer function terminals. Unfortunately, the container terminal is limited capacity and technology used transport units , thus piling up more layers to increase capacity. This increases the number of containers in the landfill , but very often difficult to locate the container , and the effective execution of transhipment operations on it. As the number of handled cargo units , there are new problems . With little turnover of the cargo container to find the company was not a problem for those involved in the physical handling of cargo. Today, however, the terminal supports more load and hence there is a possibility of a problem to locate the container that is to be assigned. During the deployment analzy intermodal cargo transshipment node can watch two indicators:

•   rotation ratio
•   the intensity of use of the component.

Rotation ratio is the number of container shifts performed per unit time with respect to one component of the terminal. This means that the residence time of the free- space component may be relatively short, but the number of containers in a given location may be large. This means that the area is heavily used, but due to the large rotation load . Achieving high turnover ratio is desirable for marine terminals, where it counts the technical efficiency of the transport process and land-based terminals , where the number of occupied seats begins to cause complications in the implementation of the basic functions of transport. The intensity of the storage is a busy time of the landfill by the same unit load per unit of time. The smaller the value of this ratio the greater the rotation of the loading unit is characterized by a terminal.

5.   DISRUPTION OF TRANSPORT OR STORAGE

5.1. Disruption in yardplanning

While making decision in yard planning system on container movement, there are three important factors that are usually taken into account before one:

•   costs of operation,
•   time of operation,
•   discrepancy, that process of transshipment will finish successfully.

Lets focus on discrepancy, as a factor close to dependability (risk) issues. This factor can be characterized as probability, that task given by yard manager will be done without delay, reliable.

There are different reasons treating different matters that should be taken into account when talking on task success. Obvesly there are some conditions prescribed to machines, its operators and tasks. Figure 1 shows simple loop of dependings in decision making and achieving success in container transshipment. Starting from 'Task success" it is assumed, that all information about machine and persons are delivered to yard planner with information on task status. This element gives opinion to yard planner about resources – technical and human. Using information from that two different sets is ready to make decisions, that treat:

- operator,
- machine.

In case of operators yard planner decisions have influence on their behavior, stress, ect. However planner's position gives opportunity to make orders, suiting operator, machine and present task condition. In case of machines, yard planner has influence on maintenance decision. The result of the decision is change in availability or reliability and maintainability performance. More details are presented on figure 2.

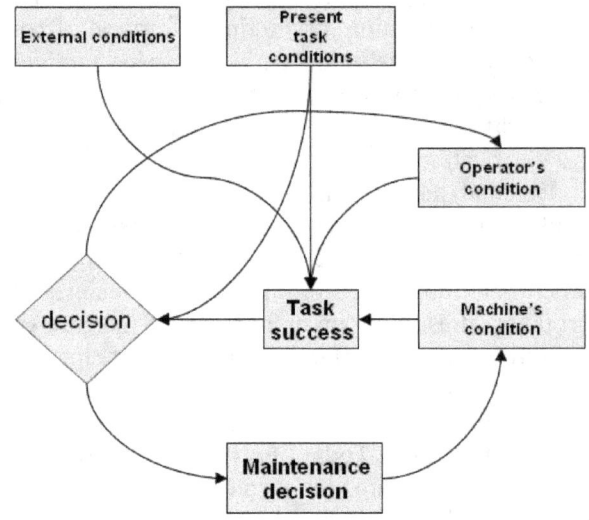

Figure 1. Fundamental relation in container yard operation

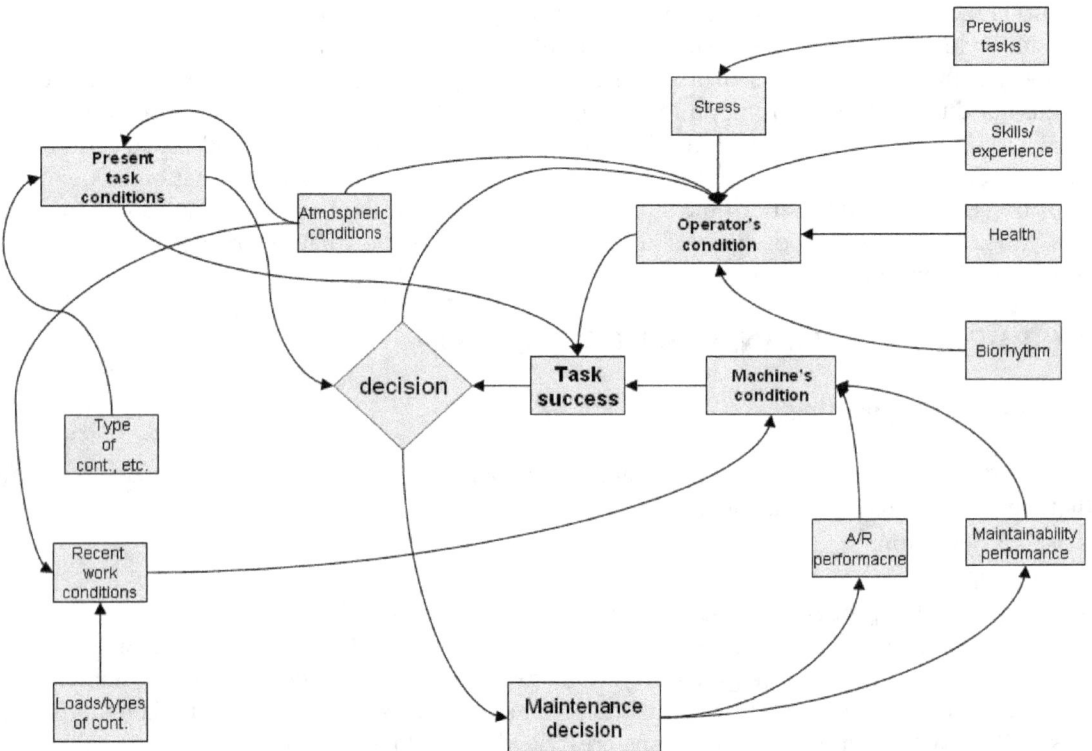

Figure 2. Extended graph of relation in container yard operation

Operator's condition is result of its skills, experience, stress or even atmospheric conditions. As it can be seen there are different modes of deciding factors, which have to be translated to common denominator. There are some open questions: how to measure stress, skills or health. Naturally there is also another question about importance of each measure/factor.

5.2. Causes and effects of disturbances

In the course of identifying the operation of intermodal transshipment node highlighted the potential for any interference during cargo delivery to the recipient. Table 1 contains the name of the fault, the cause and the possible consequences of its occurrence. Ratings shown in the table are subjective. The significance of risk determines the impact of the phenomenon on the possibility of carrying out the action with a positive result at a given time. The greater the delay resulting from the immanent danger the higher the significance of the threat. The possibility of risk reduction is an appropriate opportunity to guide the process in order to reduce the occurrence of a hazard.

Table 1 Causes and effects of disturbances during operation of the unit load. Subjective evaluation.

Description of disruption	Cause	The significance of the risk	The possibility of reducing the risk
Damage to the stage prior to the adoption of transport container terminal	Technical	high	impossible
Damage to the stage inside the terminal cargo handling	Technical	high	high
Customs Clearance	organizational	medium	high
Congestion in the course of delivery to the customer	Technical or organizational	medium	low
The problem of finding the container at the landfill	organizational	low	high
Problems with the development of transport documentation	Technical	low	high
No trailers	Technical or organizational	medium	high

As it can be seen from the table there is no practical possibility of enclosing the transport process of disruption in the event of damage to the unit load on the delivery of cargo to the terminal. But within it, and further beyond, the behavior of the relevant technical and organizational conditions can contribute to reducing the risks. Among the factors that could be somewhat overcome by the implementation of this project is to
• The problem of finding the container at the landfill,
• The problem with the development of transport documentation.

6.   CONCLUSIONS

Contemporary challenges for the transport of lead by monitoring parameters such as efficiency , productivity, quality and safety. This can be done by streamlining the demand for transport services , even within the systems already considered environmentally friendly . This will ensure that they maintain both business and cares for the environment at the same time economically rational.
The practical effect of study is to prepare a computer program to support operations in intermodal transshipment node . The developed program will support the decision of the Broadcasting space for container storage sued to obtain the following benefits:
• reducing the number of operations at the terminal intermodal,
• reducing energy needs during handling unit loads at the terminal intermodal,
• Shorter handling unit loads,
• Increased machine productivity within the terminal handling by reducing machine cycle times .

Acknowledgements

The results presented in this paper have been obtained within the project "The model of operations in intermodal terminal" (contract no. POIG.01.03.01-02-068/12 with the Polish Ministry of Science and Higher Education) in the framework of the Innovative Economy Operational Programme 2007-2013

References

[1] Ambrosino D., Caballini C., Siri S.: A mathematical model to evaluate different train loading and stacking policies in a container terminal. Maritime Economics & Logistics (2013) 15.

[2] Boysen N., Fliedner M., Jaehn F.: A Survey on Container Processing in Railway Yards. Transportation Science (2011) Volume: 47 Issue: 3.

[3] Braekers K.: Optimization of empty container movements in intermodal transport. 4or-A Quarterly Journal Of Operations Research (2011) Volume: 11.

[4] Gambardella L. M., Rizzoli A. E., Zaffalon M. (1998). Simulation and planning of an intermodal container terminal, *Simulation* 71. 1998.

[5] Olivo A., Di Francesco M., Zuddas P.: An optimization model for the inland repositioning of empty containers. Maritime Economics & Logistics (2013) Volume: 15

[6] Nowakowski T., Kwaśniowski S., Zając M.: Transport intermodalny w sieciach logistycznych. Of Wyd. PWr, Wrocław, 2008.

[7] Nowakowski T., Werbinska-Wojciechowska S.: *Means of transport maintenance processes performance : decision support system*. Carpathian Logistics Congress, CLC' 2012: congress proceedings, Jesenik, Czech Republic, November 7th-9th 2012. Ostrava : Tanger, cop. 2012.

[8] Steenken D., Voss S. Stahlbock R.: Container terminal operation and operations research - a classification and literature review, *OR Spectrum* 26. 2004.

[9] Jakubowski L.: Technologie prac ładunkowych, Warsaw University of Technology, Warszawa, 2003.

[10] Zajac P.: Can the raising of energy consumption of information interchange be a factor that reduces the total energy consumption of logistic warehouse system?, Production engineering innovations and technologies of the future, Edward Chlebus (ed.), 2011. s. 79-89, International Conference "Production Engineering 2011", Wrocław, 30 June-1 July 2011.

[11] Zajac P.: The choice of parameters of logistic warehouse system, with taking the energy into consideration, Selected logistics problems and solutions, Katarzyna Grzybowska, Paulina Golińska (eds.). Poznań : Publishing House of Poznan University of Technology, 2011. s. 107-120

[12] Vatanabe I: Contaier terminal Planning. Theoretical approach, WCN Pubishing 2005;

[13] A Quality strategy for combined transport.

[14] Combined Transport Development Strategy in Poland (in (polish), 2004 przez Generalny Departament Energii i Transportu przy Komisji Europejskiej publikacja,,*In serach of efficiency to suport intermodality growth*

[15] Vis I. F. A., de Koster R.: Transshipment of containers at a container terminal: An overview, *European Journal of Operational Research* 147, 2003.

A Risk Informed Assessment of Hydrogen Dispensing in Warehouses

Kumar Bhimavarapu[*]

FM Global, Norwood, MA, USA.

Abstract: Hydrogen dispensing units are installed in creasingly in warehouses to refuel fuel cell powered fork lift trucks. A risk inf ormed assessment was undert aken to e valuate the adequacy of safety systems with a focus on property damage from explosions resulting from accidental hydrogen releases. A few scenarios covering the potential range of releas es were evaluated. The explosion-related consequences in terms of overpressures and associated damages were taken from another modeling study. Based on failure rate data for generic and hydrogen systems, order of magnitude likelihoods were assessed fo r the release and explosion scenarios. The estimated property damage risk was evaluated against tolerable ri sk established using three independent criteria based o n severity of consequence s, a SIL (Safety Integrity Level) matrix, and loss experience in warehouses. Risk reduction opportunities were identified in terms of the integrity of the safety functions performed by the instrumentation.

Keywords: hydrogen dispensing, explosion hazard, warehouses, risk informed assessment

1. INTRODUCTION

As part of "green" solutions, hydrogen is becoming popular as a fuel in fuel cell driven systems. Fuel cell powered forklift trucks are increasingly used in warehouses. Loss experience with these sy stems thus far has been favorable. However, the practice of indo or dispensing of hydrogen introduces new fire and explosion hazards to the t ypical warehouse occupancy. Therefore, it is critical to understand and manage the associated property risks.

A detailed literature survey indicated t hat none of the codes and standards or published studies dealt specifically with the property damage and the associated risk from the hydrogen release and the consequent explosion in l arge enclosed spaces such as warehouses. In this connection, a study was undertaken to evaluate the hazard from hydrogen dispensing operation in l arge warehouses and the associated risk from property damage considerations, and to identify the risk reduction opportunities.

2. DISPENSING SYSTEMS

There is a variety of process design options for disp ensing gaseous hy drogen. The fueling installations that we reviewed ty pically include a bu lk liquid hydrogen storage, com pressors and tanks for gaseous hydrogen, all of which are located outside the buildi ng. Gaseous hydrogen is hard-piped into the building where dispensers are generally located along t he interior perimeter wall. Self-service hydrogen f uel-dispensing systems include key, code and card lock systems, which allow filling of permanently mounted fuel containers on hydrogen-powered vehicles.

3. SCOPE OF THE ANALYSIS

Standards such as NFPA 2 (2011) [1] and N FPA 55 (2013) [2] provide guidance on process and engineering safety requirements, and safety functions that nee d to be addressed in the design and operation of a h ydrogen dispensing system. The main safety functions are related to p revention of dispensing in case of an abnorm al condition, and shut down and isolation of the dispenser in case of a hazardous condition. In addition to the sensors th at monitor system variables such as pressures,

---

[*] Email:Kumar.Bhimavarapu@FMGlobal.com

temperatures, and flow rates, gas and fire detectors are provided to shut down and isolate the system in case of a hazardous situation.

The process design and safety instrumentation change with the process. Accordingly, process specific hazard analyses are needed to ensure that the process and safety instrumentation are adequate to address the needed safety functions. The analysis presented in this paper considers that the process design (including process and safety instrumentation) and operating procedures commensurate with good engineering practices and applicable codes and standards, and are subjected to appropriate hazard analysis. The focus is on identification of the integrity requirements of the needed safety functions based on the likelihood and severity of the consequences from hydrogen related explosions and fires. Identification and evaluation of engineering safety requirements and safety functions are not delved into in this paper.

4. SCENARIOS OF CONCERN

By considering that the available instrumented and safety systems will be adequate to address the needed safety functions, the initiating events of concern are releases of hydrogen from mechanical integrity failures of components of the dispensing system. No operator errors (such as opening of valves by mistake) that alone can lead to unmitigated releases are envisaged in the dispensing process, which is controlled by automated PLC-based instrumentation.

As long as the instrumented systems function per the design intent, hydrogen releases are either prevented or minimized, and no credible loss is expected. The mitigated scenarios are not of serious concern from a property loss exposure perspective. When instrumented systems fail to isolate the hydrogen supply in case of a hazardous condition, hydrogen continues to release until an action is taken or the system is emptied out. The focus of the analysis is on such scenarios where uncontrolled hydrogen releases are ignited leading to explosions and fires that can result in significant damage.

4.1 Mechanical Integrity Failures

A schematic of the type of hydrogen dispensing unit considered in this study is shown in Figure 1. The main header into the building connects to the individual sections of piping that supply hydrogen to the two dispensers operating at different pressures, viz., 25MPa and 35 MPa. In order to limit the maximum flow in case of an accidental release, restricted orifices (ROs) are provided on hydrogen piping: one on the main piping header outside the warehouse (RO1), and one on each supply header inside the warehouse to the dispensers (RO2 and RO3).

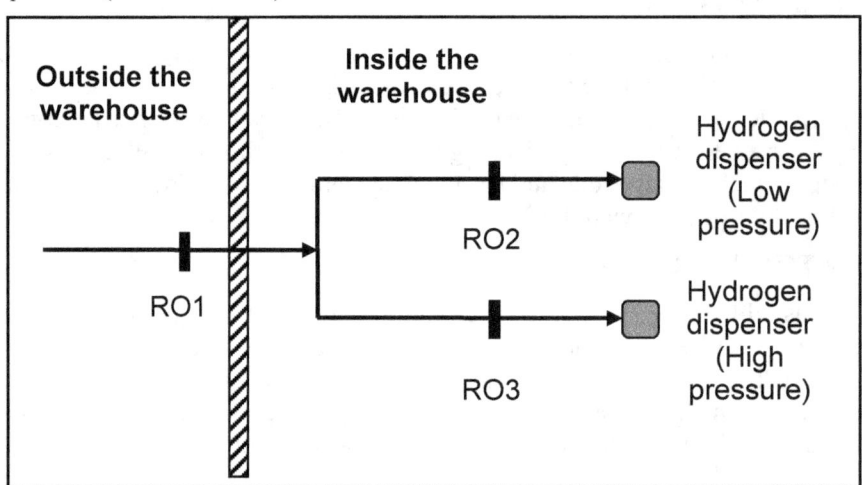

Figure 1: A Schematic of Hydrogen Supply to Hydrogen Dispensers

The dispensing system inside the building is comprised of a few meters of 1/2 - 3/8 inch diameter piping, appurtenances (such as valves and instrumentation), and hoses. The dispenser nozzle is a mechanical double-block-bleed valve that is independent of the electrical controls. In case of a pullout of the hose, the valves at the inlet of the hose (break-away valves) close automatically.

It should be noted that, based on the piping diameters (1/2 to 3/8 inch) and the pressures under consideration, in case of a "guillotine" type break of piping inside the warehouse, the flow rates could be up 20 to 30 kg/min, even though the rate can decrease over time considering the nature of the system and the inventory under consideration. However with ROs in place, such flow rates are reduced considerably, at the same time meeting the functional requirements of the dispenser. Based on the ROs considered in these evaluations, in general any break in piping downstream of the ROs on individual supply headers will release a maximum flow of 2 kg/min. While the ROs in the headers to the dispenser are expected to limit the flow, the exact flow would be dependent on the site specific design, and the size of the breach.

4.2 Releases

Assessment of consequences needs evaluation of a variety of releases, and consequent scenarios. The likelihoods of release, release rate, and duration of the release change with the scenario. Five constant release rates, viz., 0.25, 0.5, 1, 2, and 4[1] kg/min were examined to cover the range of scenarios of concern and interest.

The release duration considered in this study is 3 minutes - a moderate value. The warehouses of interest are attended 24 hours per day and 365 days per year. The remote and manual system shutdown switches recommended by codes and good engineering practices are expected to be readily accessible to personnel nearby. Sometimes, a timer facility is provided in the dispenser. Based on the available designs of the fuel cells, a 3-minute duration is considered adequate for dispensing the required quantity of hydrogen. Considering the nature of the warehouses and the dispersion of vapor cloud, an ignition is expected by 3 minutes after the release starts. In case of an ignition, further accumulation of hydrogen is not considered credible.

4.3 Ventilation

Typically, the amount of ventilation during a given release time is fairly small. Thus normal building ventilation would have little mitigating effect on dispersion and amount of flammable material available in the vapor cloud and in turn on the severity of explosion in general. However, the ventilation could effectively provide a degree of explosion venting since pressure buildups are slow during deflagrations in such large warehouses with relatively slow flames. Detonations, however, cannot be vented since the pressure increases so rapidly that the vent opening has no impact on the maximum pressure. Two cases were considered to evaluate the effect of ventilation for various release scenarios.

Case 1- Minimal/No Ventilation: Some warehouses may not have any designated mechanical ventilation.

Case 2 - Three air changes/hour: Ventilation rates change based on the climate and demands. Based on the observed ventilation rates at various warehouses, a ventilation rate of 3 air changes/hr was chosen in our evaluation as the second case.

5. EXPLOSION ANALYSIS

In order to generate the needed understanding of the consequences of potential explosions from uncontrolled releases of hydrogen, realistic modeling is needed for dispersion and explosion in large enclosed places. The results of a CFD modeling study undertaken by Bauwens (2013) [3] to evaluate

---

[1] Even though the release rate of 4 kg/min was evaluated, based on the ROs considered in the design, it is not a possible release scenario.

hydrogen explosions in large wareho uses were used in this analy sis. The details of the basis of the modeling and the generated results used in the current study are described below briefly.

5.1 Basis of Modeling

A warehouse of size 62.4 m (W) x 62.4 m (L) x 8 m (H) was considered adequate to represent th e warehouses of interest and to evaluate the phy sical effects of hydrogen release, dispersion and deflagration. The number and the height of racks considered were those of a typical warehouse.

The hydrogen dispenser was located along the wall of the warehouse a few meters from one of th e corners. A corner provides the most conservative release location within the warehouse. If the rele ase occurs away from the corner of the warehouse then the cloud expands in more directions and results in a thinner cloud with more mixing, and a lower mass above the LFL.

No obstructions were considered on the ceiling. Further, in order t o create a conservative estimate of the amount of released hy drogen in the flammabl e mixture, it was assumed that the hydrogen r eleased from the dispenser hits a solid surface, rel eases with low momentum and rises as a buoy ant plume. The simulations were run for three-minute releases. The final mass of hy drogen above the LFL at the end of 3 minutes was used to estimate the maximum overpressure.

5.2 Results of Explosion Modeling

Overall, the consequences were observed to decrease with an increase in the size of the warehouse and the availability of ventilation.

5.2.1 Overpressures from Explosions

For all of the evaluated re leases, the main source of da maging overpressure i s the slow combustion of hydrogen, which without venting results in pressuriza tion of the warehouse. In explosions, dam age may also occur d ue to fla me acceleration and the generation of a blast w ave; however, for the releases evaluated in this study, the damage caused by the blast wave was estimated to be minor. The overpressure results were estimated for two cases: i) when only the mass of hydrogen above the LFL is consumed (P_{max} above LFL), and ii) when all of th e hydrogen released is consumed (P_{max} Total). Table 1 summa rizes the results of the dispersion sim ulations as well as the o verpressure estimates generated for different rele ase rates.

Table 1: Full Scale Unventilated Warehouse Simulation Results [3]

Release Rate (kg/min)	Total Mass Released (kg)	Total Mass Above LFL (kg)	Mass above LFL	P_{max} above LFL (bar/psi)	P_{max} Total (bar/psi)	Blast Wave Radius (m)
0.25	0.8	0.03	4.1%	-	0.01/0.145	-
0.5	1.5	0.34	22%	0.01/0.145	0.02/0.29	-
1	3.1	1.6	52%	0.02/0.29	0.04/0.58	5
2	6.1	4.1	67%	0.06/0.87	0.09/1.3	13
4	12.4	8.0	64%	0.12/1.74	0.18/2.6	23

The results presented in Table 1 are for the no ventilation case, i.e., the enclosure is well sealed without venting. In addition to the overpressur e results, the table also includes estimates of the radius withi n which light damage may occur due to the generation of a blast wave. Blast wave estimations were done using an FM Global proprietary software package.

The effect of ventilation was examined by extrapolating peak overpressure results (considering only the mass above the LFL) with ventilation (three air changes per hour) and without ventilation, and also for variations in the warehouse size. Table 2 shows extrapolated peak overpressure results.

In addition to the pressure damage consequences of a hydrogen explosion, the possibility of unintended sprinkler activation due to the deflagration was also examined. Based on estimates, in some cases, it was found that all sprinkler heads located within the flammable cloud would likely activate following a hydrogen deflagration.

Table 2: Peak Overpressures Estimated for Additional Warehouse Geometries [3]

Height (m)	Area (m^2)	Volume (m^3)	1 kg/min		2 kg/min		4 kg/min	
			Closed (bar)	Ventilated (bar)	Closed (bar)	Ventilated (bar)	Closed (bar)	Ventilated (bar)
8	1950	15600	0.046	0.029	0.122	0.085	0.243	0.211
8	3900	31200	0.023	0.007	0.061	0.024	0.119	0.083
8	7800	62400	0.012	-	0.030	0.001	0.059	0.024
12.8	1950	24960	0.029	0.012	0.076	0.038	0.150	0.114
12.8	3900	49920	0.014	0.001	0.038	0.005	0.074	0.038
12.8	7800	99840	0.007	-	0.019	-	0.037	0.006

6. JET FIRE ANALYSIS

From a property loss exposure perspective, all hydrogen related fires (other than the fire following an explosion) are considered as ignition sources. SuperChems[2], a commercial software package was used to evaluate jet dispersions and jet fires.

The analysis also reviewed the potential overload on sprinkler systems from an ignition of combustibles initiated by jet fires. Based on the estimated size of the jet fires, they are not expected to result in large losses of concern on their own, provided adequate distances are maintained between the dispenser and the combustible storage.

7. LIKELIHOOD EVALUATION APPROACH

The likelihood of an explosion is a function of i) likelihood of accidental release of hydrogen; ii) failure of the available instrumented system to shut down and isolate the system to stop the release; and iii) failure of human intervention if the instrumented system fails. Since the perils of concern are fires and explosions, ignition likelihood and likelihood of a deflagration in case of an ignition are also important variables.

7.1 Release Likelihoods

After a detailed review of the compiled leak frequency data from literature, data from two studies by LaChance (2009) [4] and HSE (2012) [5] were mainly used in this study as they are relevant to the hydrogen dispensing systems under consideration.

LaChance (2009) [4] presents the results of a study performed to support the development of risk-informed[3] separation distances with focus on gaseous hydrogen storage facilities and the impact on the public at large. From the generic leakage frequency data compiled for various components, LaChance

---

[2] 'SuperChems[TM] Expert' is a commercial software package from ioMosaic Corporation.

[3] The risk under consideration in this study is related to personnel safety from jet fires.

(2009) [4] estimated hydrogen specific leakage frequencies using Bayesian analysis. Even though the focus was on personnel safety and outdoor facilities, the leakage frequency data developed by LaChance (2009) [4] was considered applicable for hydrogen dispensing units in enclosed spaces as well. LaChance (2009) [4] provides hydrogen release frequency data (median and 95% confidence values) for very small, minor, medium, major, and rupture leak sizes which correspond to leak areas of 0.01%, 0.1%, 1%, 10%, and 100% of total flow area respectively. Overall, the leak frequencies presented for hydrogen systems are one to two or even three orders of magnitude less than the compiled generic data. For larger leak sizes, hydrogen release estimates are much closer to general frequencies (by a factor of 2 to 4) as compared to smaller leak sizes. For hoses, there is a considerable difference between generic and hydrogen system leak frequencies.

In the dispensing system of our interest, even though piping and appurtenances are 1/2 to 3/8 inch size, maximum flows are restricted irrespective of the size of the breach in view of the presence of ROs in the piping. Any release from a rupture (100% leak area) to a medium leak (1% of flow area) in the piping or appurtenance is of concern since it can lead to an explosion and property damage. Thus the release likelihood values estimated are the sum of the frequencies of medium, major and rupture leakages. Minor and small leak frequencies are not included since minimal damage is expected from those scenarios.

Components of the dispensing unit being evaluated in this study are only those located in the building and include a few meters of piping, appurtenances, and hoses. The appurtenances considered include main isolation valve, control/shutoff valves on the main and supply headers, filters, restricted orifices, pressure control valve and relief valves on the low pressure header, and instrumentation connections. In order to be conservative, 95% confidence values were used for leak frequencies.

HSE (2012) [5] provides data on failure of hoses (leading to releases) with and without mitigation systems including break-away valves, reportedly collected from chlorine facilities. The H_2 incident database [6] reports around 19 incidents associated with hydrogen fueling stations. Even though the details are not available, three of them are associated with break-away connections. The break-away designs of a hydrogen hose are supposed to be more robust than that of a typical hose; however, based on the available data and our experience, high integrities couldn't be assigned to break-away valves. Using the data provided in HSE (2012) [5] and the loss experience, a range was considered for the failure likelihood of hoses and break-away valves.

Keeping in view the potential variations in the length of piping, and number of appurtenances and hoses from location to location, parametric estimations were performed and ranges of likelihoods were estimated.

7.2 Other Likelihoods

Hydrogen's flammability range (between 4% and 75% in air) is very wide compared to other fuels. Under optimal stoichiometric combustion condition, the energy required to initiate hydrogen combustion is much lower than that required for other common fuels. Tchouvelev (2008) [7] reported the ignition probabilities estimated for hydrogen by considering the available ignition probability data of other fuels, and their properties as compared to hydrogen. The ignition likelihood value provided by Tchouvelev (2008) [7] for enclosed areas was chosen considering that the warehouses are also enclosed spaces, and may not have any special ignition control systems.

Based on the observed trend in the industry, instrumented systems meant to address safety functions in dispensing systems are not considered as certified safety instrumented systems (SISs). Accordingly the instrumentation serving safety functions was not considered to provide even one order of magnitude (a factor of 10) risk reduction.

Considering the limitations in the available data due to the limited experience of the industry with hydrogen dispensing units, we recognize the uncertainty associated with various likelihood values chosen and estimated. Accordingly only order-of-magnitude likelihood values were used for our estimations.

8. ADEQUACY OF SAFETY ASSESSMENT

In order to assess the adequacy of safety systems, in addition to identification of safety functions, it is necessary to identify the needed and the available safety integrity of those safety functions. With the advent of PLC based controls, there is a need and opportunity to define the required integrity of the safety functions and also to evaluate the available integrity in the instrumentation serving the safety functions. SIL (Safety Integrity Level)[4] requirements for the safety functions depend on the process, instrumentation, operation, the material being handled and, importantly, the tolerable risk.

In view of the imprecise nature of the estimations, three independent approaches were used to establish tolerable risk and assess the adequacy of the available safety systems[5]. The first approach is based on the severity of consequences alone. The remaining two are risk informed approaches, where likelihoods are considered in addition to consequences. The second approach involved the use of a SIL (Safety Integrity Level) matrix. Based on the likelihood of the initiating event and the severity of the consequences of the scenario, SIL matrices are used to identify the needed safety integrity in terms of SILs for the safety functions to be performed by the instrumented systems. The third approach is based on the premise that loss experience can form the basis for the tolerable risk since such loss exposure has been tolerated by industry. An f-N curve generated from loss data provides the cumulative frequency of losses exceeding any specific loss $ value and thus the acceptable likelihood of a scenario or scenarios with greater than a specific property damage value.

In order to facilitate the implementation of the three above mentioned approaches, the needed data[6] were generated using the methodologies outlined in the previous section and FM Global's in-house data on warehouse related losses and property values. Warehouses were grouped into three size-based categories: small ($<50,000$ m^3), medium ($50,000$ to $100,000$ m^3) and large ($>100,000$ m^3). The estimated overpressures as a function of the size of the warehouse were used to identify the extent of damages for the selected release scenarios for the three categories of warehouses. The effect of potential sprinkler system failure due to explosion was also included in property damage estimations. Typical property values per square footage and the associated business interruption values were used for these estimations.

8.1 Severity-based Approach

Table 3 provides a qualitative assignment of the severity of impact of over pressures as a function of the size of the warehouse, the presence of ventilation, and release rate of hydrogen. Table 4 presents the description of color coding that is used in Table 3.

8.2 SIL Matrix-based Approach

Figure 2 provides the SIL (Safety Integrity Level) matrix used in this study to identify the SIL rating of the safety function to be fulfilled by the safety instrumented systems (SISs). Tables 5 and 6 provide the descriptions of the probability and consequence categories used in Figure 2.

The SIL rating is based on the likelihood of an initiating event, the severity of the consequences, and the number of available non-SIS IPLs (Independent Protection Layers).

---

[4] SIL (Safety Integrity Level) indicates the order of magnitude of risk reduction provided by the safety functions implemented in safety instrumented systems (SISs).

[5] Safety systems of interest in this study are the safeguards or mechanisms to prevent/minimize property damage and business interruption related losses.

[6] The in-house data and estimated numbers are not presented in this paper; however, the methodology and the conclusions drawn are presented for the benefit of the industry.

Table 3: Effect of the Size of the Warehouse on Extent of Damage

Release rate, kg/min	0.5	1	2	4	0.5	1	2	4
Continuous Ventilation	None	None	None	None	Present	Present	Present	Present
Warehouse size	Severity of overpressure							
Small warehouse								
Medium								
Large								

Table 4: Severity Rating for Consequences

Qualitative Severity rating	Color code used in Table 3
No concern	
Negligible concern	
Low impact	
Medium impact	
Significant impact	

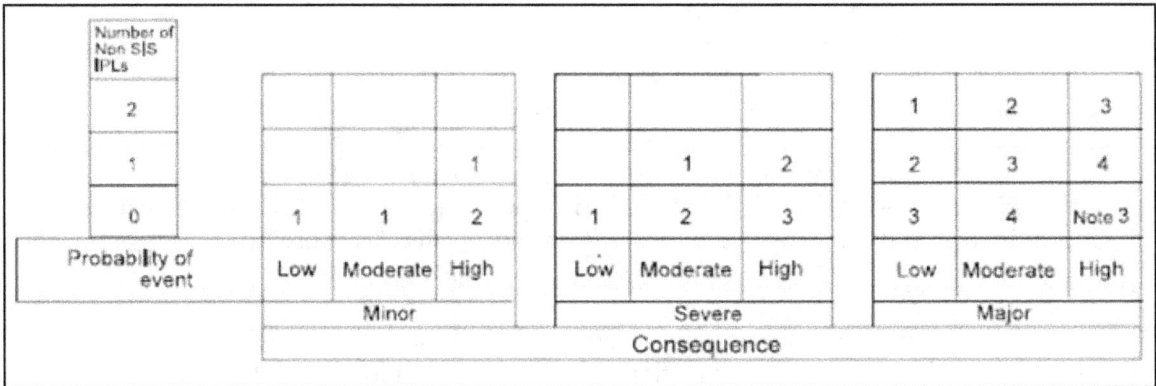

Figure 2: SIL Matrix [8]

Table 5: Consequence—Typical Categories [8]

Category	Description
Major	Substantial or total destruction beyond a local process area into the surrounding plant and shut down more than 3 months
Severe	Substantial damage mainly to a single process area and shutdown from 2 weeks to 3 months
Minor	Equipment damage and shutdown less than 2 weeks

8.3 Loss Experience-based Approach

An f-N curve such as shown in Fi gure 3, generate d based on warehouse loss e xperience, was used to identify the exceedance frequencies for loss $ value of the property damage of interest.

Table 6: Probability of Process Upsets—Typical Categories [8]

Probability	Type of Initiating Event
High	Failure can be expected within the life of the plant. *Examples include process leaks, single instrument or valve failures, or human errors that result in releases of hazardous material*
Moderate	Failure or series of failures may occur with low probability within the life of the plant. *Examples include dual instrument or valve failures, a combination of instrument failure and human error or large releases in loading/unloading areas.*
Low*	Failure may occur with very low probability within the life of the plant. *Examples include combinations of multiple instrument failures and multiple human errors, or full-bore failures of small process lines or fittings.*

Note: * Multiple instrument or valve failure, multiple human errors, or spontaneous failures of vessels have lower likelihood but may also be considered in this group.

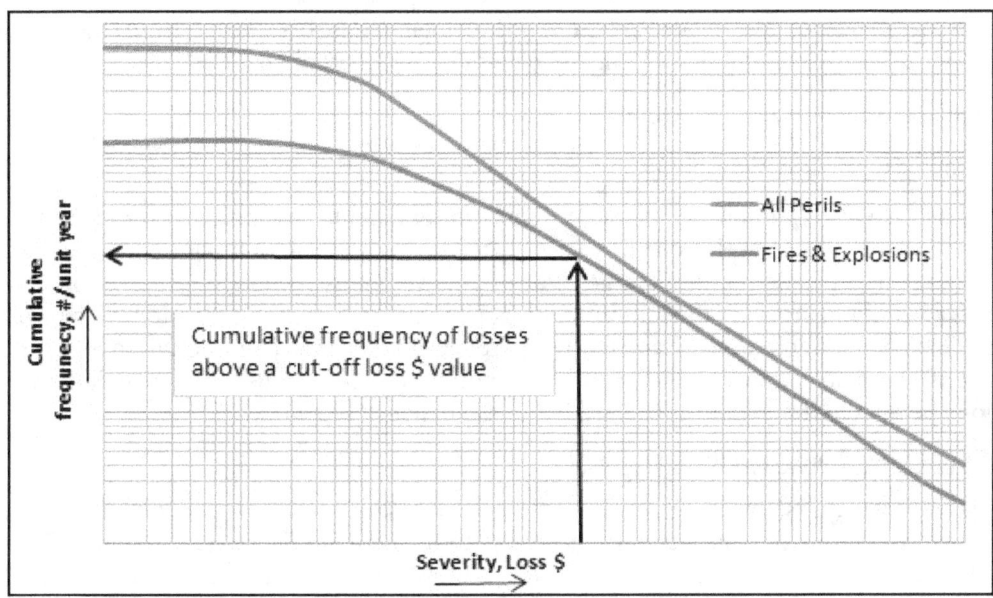

Figure 3: f-N Curve for all Perils and Fires & Explosions Losses

9. CONCLUSIONS

The conclusions drawn based on the estimated tolerable risk criteria, likelihoods and severities of explosions discussed in previous sections are provided in this section. Overall, hydrogen jet fires were not considered to affect the risk (property damage-based) at the estimated resolutions.

At this time, due to the limited operating experience on hydrogen dispensing systems, uncertainty is associated with the chosen likelihood values, and also the consequences considering the possible variations in release duration, nature of dispersion, and severity of the fire following explosion. However, the following risk reduction opportunities (RROs) are considered appropriate until significant experience is accumulated and data are collected.

9.1 Integrity of the Safety Functions

The risk reduction opportunities (RROs) related to the safety integrities of the instrumentation that serve safety functions are as follows. The following RROs assume that the instrumented safety functions provided in the design are adequate.

a. Based on the possible damage, for small (<50,000 m^3) warehouses without ventilation, it is necessary to have an assured SIL1[7] integrity for the safety functions performed by the instrumentation.

In case of a hazardous condition, safety functions i) prevent the startup of the dispensing system and ii) shut down and isolate the dispensing system if in operation.

b. For medium (50,000 – 100,000 m^3) warehouses without ventilation and for small (<50,000 m^3) warehouses with ventilation, the damage potential is present, and thus it is desirable to have an assured SIL1 integrity for the safety functions performed by the instrumentation as mentioned above.

c. For the remaining warehouses, no certified safety instrumented systems are considered necessary to perform the safety functions.

If confirming the needed integrity in the PLC-based instrumentation is not feasible for any reason, dedicated hardwired Safety Instrumented System/s (SIS/s) may be considered to ensure the needed safety integrity. Even though gas detection and fire detection appear to be a separate layer of protection, generally they are connected to the same final control element (shutoff valve) to fulfill the safety function requirement; hence, they are not independent and cannot be given credit as IPLs.

9.2 Other Risk Reduction Opportunities (RROs)

9.2.1 Provision of Dedicated Ventilation System

Even at a release rate of 1 kg/min, the potential exists for damaging overpressures, and further means of mitigating the hazard such as explosion venting, or direct exhaust ventilation above the hydrogen dispenser could help. Considering the large size of the warehouses under consideration, we did not evaluate higher ventilation rates in the building as a potential solution. However, it may be worthwhile to evaluate spot/local ventilation directly above the hydrogen dispenser as an option to minimize the hazard associated with explosions. In such cases, the ventilation rate must be sufficient to keep the concentrations below the flammable concentration in the exhaust ductwork.

9.2.2 Maximum Flow Rates into the Building

If hydrogen dispensing systems have to be located indoors, the maximum hydrogen flows into the building should be controlled to the extent possible with the help of restricted orifices (ROs) inside and outside the building. If the allowed flow rates exceed the 2 kg/min value used in this study, additional evaluations and/or measures may be needed to address the risk.

9.2.3 Hydrogen Dispensing Loss/Failure Data

Gasoline and diesel dispensing has been in use for a long time, and this has resulted in codes and standards that have yielded an acceptable loss record. Even though hydrogen dispensing systems are more hazardous and the requirements may differ from indoor to outdoor dispensing systems, with adequate instrumentation, training, and administrative controls, hydrogen dispensing could also be implemented with an acceptable loss record. Accumulation of data is an important task in order to establish a quality basis for estimating hydrogen release frequencies and accordingly the needed layers of protection and their integrity.

ACKNOWLEDGEMENTS

The author acknowledges the support provided by Regis Bauwens of FM Global for indoor dispersion and explosion modeling; and the support provided by Amy Brown and Glenn Mahnken of FM Global for their technical input in choosing the scenarios and associated parameters, and technical review of the analysis.

---

[7] SIL1 (Safety Integrity Level1) provides one order of magnitude of risk reduction.

REFERENCES

[1] NFPA Standard 2: Hydrogen Technologies Code, NFPA, 2011.

[2] NFPA Standard 55: Compressed Gases and Cryogenic Fluids Code, NFPA, 2013.

[3] Bauwens, C.R., and S.B. Dorofeev, CFD Mode ling and Co nsequence Analysis of an Accidental Hydrogen Release in a Large Scale Facility", Proceedings of the 5th International Conference on Hydrogen Safety, Brussels Belgium, 2013.

[4] LaChance, J., W. Houf, B. Middleton, and L. Fluer, Analyses to Support Development of Risk-informed Separation Distances for Hy drogen Codes and Standards. SAND2009-0874, Sandia National Laboratories, March 2009

[5] Health and Safety Executive, Failure Rate and Event Data for Use within Risk Assessments, UK, 2012.

[6] H_2 Incident Database, web-based, 2012, http://www.h2incidents.org/

[7] Tchouvelev, V. A., Knowledge Gaps in Hy drogen Safety; Main Report – Survey of Hydrogen Risk Assessment Methods, A. V. Tchou velev and Asso ciates, Report No.: 2 005-1621 Rev 2, 2008, DNV Research & Innovation.

[8] FM Global Loss Prevention Data Sh eet 7-45: Instrumentation and Control in Safety Applications, FM Global, 2000.